D0096107

HOW TO DEFEAT YOUR OWN
CLONE

And Other Tips for Surviving
the Biotech Revolution

KYLE KURPINSKI
AND
TERRY D. JOHNSON

BANTAM BOOKS TRADE PAPERBACKS
NEW YORK

Kyle and Terry would like to dedicate this text to their respective parents
for their patience, support, and genetic material

A Bantam Books Trade Paperback Original

Copyright © 2010 by Kyle Kurpinski and Terry D. Johnson

Published in the United States by Bantam Books,
an imprint of The Random House Publishing Group,
a division of Random House, Inc., New York.

BANTAM BOOKS and the rooster colophon are registered trademarks of
Random House, Inc.

Library of Congress Cataloging-in-Publication Data
Kurpinski, Kyle.
How to defeat your own clone : and other tips for surviving the biotech revolution /
Kyle Kurpinski and Terry Johnson.
p.   cm.
ISBN 978-0-553-38578-6
1. Biotechnology—Popular works.   2. Human cloning—Popular works.
I. Johnson, Terry.   II. Title.
TP248.215.K87 2010
660.6—dc22   2009045899

Printed in the United States of America

www.bantamdell.com

2 4 6 8 9 7 5 3

Book design by Casey Hampton

Illustrations by Ming Doyle

CONTENTS

PROLOGUE

Cloning has got to be the dumbest idea in history. Have scientists never watched a single movie in their lives?
—STEPHEN COLBERT, *I AM AMERICA (AND SO CAN YOU!)*

You have two choices.

You could doom yourself to a future in which you compete against your own clone while your body and mind slowly degrade under assault from terminal ailments and encroaching entropy.

Or you could buy this book. We don't guarantee it'll save your behind, but it couldn't hurt. Besides, there are pictures.

In the not-so-distant bioengineered future, you'll be faced with a lot of awesome possibilities—like genetic face-lifts and chocolate-flavored broccoli. Those reaping the benefits of the biotech revolution will be healthier, smarter, and longer-lived than those who don't.

Along with these benefits will come a lot of scary possibilities—like viral warfare and biologically enhanced Richard Simmons clones. It's the scary stuff that you'll need to watch out for, and knowledge will be your best line of defense, because the biotech revolution is about more than just copying sheep; it's about reengineering life from the ground up: both yours and your genetic counterpart's.

To prepare you for a world where your genes can be deleted, rewritten, or even copied in whole, this book will teach you how your body works, how it doesn't work, how it might work in the future, and what to do if it gets cloned and the clone starts screwing with your life. Think you can outsmart yourself? Think you can beat yourself in a fight? Think it can't happen to you? Think again, bub.

The first rule of cloning is: Don't ever let your clone read this book.

HOW TO DEFEAT YOUR OWN CLONE

ONE

THE BIOTECH REVOLUTION
(Engineering Life for Fun and Profit)

In these days, a man who says a thing cannot be done is quite apt to be interrupted by some idiot doing it.

—ELBERT GREEN HUBBARD

Any sufficiently advanced technology is indistinguishable from magic.   —ARTHUR C. CLARKE

et's face it: cloning and genetic manipulations can be kinda scary. But *why* exactly? Is it because biotechnology is inherently dangerous, or is it because we don't fully comprehend it? The history of our technological development can be viewed as a series of discoveries in tension with human culture. Few innovations are welcomed with unrestrained admiration, especially when they upset the established social order. Take for example the Luddites—a British social movement opposed to the innovations brought about by the Industrial Revolution—who destroyed wool and cotton mills because these worked faster and cheaper than people did. Or consider how the splitting of the atom has changed the faces of warfare, diplomacy, and technology forever. It's understandable that people are nervous about change, especially when said change involves unemployment or hiding under desks in the event of nuclear catastrophe.

We do tend to forget that there are many innovations that once caused us great anxiety, but have come to be accepted as a part of modern life. In ancient Greece the airless space of a vacuum was a philosophical conundrum, mysteriously capable of extinguishing flames and causing small animals to die. Nowadays, it's a convenient way to get crumbs out of the carpet.

Not infrequently, the answer to an innovation's dangers is more innovation. When human beings first started to congregate in large cities, disease grew to be such a problem that there was serious speculation that living in large cities was unnatural and unavoidably dangerous. People were not meant to live so close to one another. Cities were a disastrous and doomed experiment in living!

Then plumbing happened.

Today, critics assert that biotechnology is inherently dangerous. They argue that stem cell therapies will cause cancer, bioterrorism will be the downfall of the human race, and cloning is just a way of "playing God." All the while transhumanists (those who wish to improve humanity via technology) are telling us the exact opposite: stem cells are the key to unlimited healing and virtual immortality, bioengineered viruses will cure disease, and genetic engineering is just body modification on the molecular level. Those few scientists who make their opinions known outside of impenetrable academic journals tend toward cautious optimism, while the politicians who regulate these breakthroughs and the media who disseminate them into popular culture are currently in a contest to prove which can be more flagrantly ignorant. It's a dead heat.

In the long run, it doesn't really matter what any of these groups say or believe. Progress is inevitable. Whether you love it, hate it, or simply do your best to ignore it, the biotech revolution has begun. We can't tell you exactly what the future holds—only prophets or madmen truly know the future, and we authors are overqualified for one of those positions and underqualified for the other. So, we place our reputations in a certain amount of peril, making moderately educated guesses about the biology and medicine of tomorrow. We did this for two reasons: someone had to, and we needed the money.

One final warning before we begin. This book will treat you like a machine. For those who find this dehumanizing, please keep in mind that we think you are an absolutely marvelous, unbelievably fascinating machine that we have devoted our professional lives to understanding. If you want to prepare yourself for genetic body upgrades and outrageous clone battles, it's best to consider the engineering advantages of our point of view.

A BRIEF HISTORY OF BIOTECHNOLOGY

Isaac Newton—one of history's greatest scientists—is often quoted as having said, "If I have seen a little further, it is only by standing on the shoulders of giants." (Considering what Newton thought of some of his contemporaries, however, he might have been wearing cleats.) Before we clamber upon our betters for a peek at the future, let's take a look at what has already come to pass.

- Approximately 15,000 B.C.E.—*Homo sapiens* domesticate their first animal, *Canis lupus familiaris*, the subspecies of wolf we call dogs. Having acquired a best friend, the human race decides to see if there's anything else in nature that could use a bit of tweaking.
- Approximately 10,000 B.C.E.—The sowing and harvesting of plants begins in the Middle East. This quickly replaces the previous method of acquiring food from the earth, which is best described as "wander around a lot and avoid those little red berries." Farmers are soon selecting crops for specific characteristics such as yield, resistance to disease, and deliciousness.
- Approximately 8,000 B.C.E.—The cow is domesticated and eventually becomes the Swiss Army knife of agriculture, driving plows and providing meat, milk, leather, and manure.
- Approximately 4,000 B.C.E.—After several thousand odorous years of riding in carts behind cows, human be-

ings domesticate the horse. Someday, we'll pull our carts with giant, genetically engineered lobsters, but let's take things one step at a time.

· Approximately 400 B.C.E.—The ancient Greeks begin writing seriously about biology and medicine, setting down principles that will confuse and mislead scientists for centuries.

· 1865—Gregor Mendel reads his paper "Experiments in Plant Hybridization" at two meetings of the Natural History Society of Brünn, Moravia. Initial response is underwhelming, but the scientific mainstream eventually recognizes him as "the Father of Modern Genetics," thanks to his groundbreaking work with pea plant heredity, shedding new light on how biological traits are passed to successive generations. Our understanding of pea biology quickly surpasses that of all other legumes.

· 1902—Walter Sutton and Theodor Boveri independently propose that chromosomes may be the basis for Mendelian inheritance. Minds are blown.

· 1909—Wilhelm Johannsen coins the word "gene" to describe Mendel's fundamental unit of heredity. This single careless decision leads to a century of irritating puns comparing genetics to denim pants.

· 1928—Hans Spemann performs the first transfer of nuclear genetic material in amphibians, establishing the fundamental basis of laboratory cloning.

· 1944—Oswald Avery demonstrates that DNA is the genetic material of the cell.

- 1952—Robert Briggs and Thomas J. King perform the first successful animal cloning from early embryonic cells with northern leopard frogs. Unfortunately, this is just a specific species of frog, not a frog/leopard hybrid, but yes, that would be sweet. Give us a few years.

- 1953—James Watson and Francis Crick publish their double-helix structure of DNA, with some underappreciated help from Rosalind Franklin's experimental data (the lousy sexist fifties).

- 1958—F. C. Steward grows a complete carrot plant from a single cell taken from the root of an adult plant, demonstrating that it's possible to create a clone of at least one organism from an adult cell. Steward spends the rest of his career politely pretending that he's never heard the "What's up, Doc?" joke before.

- 1963—J.B.S. Haldane coins the term "clone," without which we would have to call this book *How to Defeat Your Own Genetically Identical Human*. Thanks for the snappy title, J.

- 1966—Marshall Nirenberg and Heinrich J. Matthaei decipher the genetic code, which turns out to be much better written than *The Da Vinci Code*.

- 1972—Paul Berg creates the first recombinant DNA molecules, containing DNA sequences that are not normally found together. This technique becomes known as "gene splicing," and is fundamental to genetic manipulation and engineering.

- 1972—Walter Fiers sequences the first gene, for a protein

that forms part of the coating on a virus. The virus immediately begins questioning its own identity and self-determination.

- 1973—Stanley Cohen and Herbert Boyer create the first recombinant DNA organism—transgenic *E. coli* bacteria that contain a DNA sequence from frogs. *E. coli* has since become the go-to organism for producing large amounts of relatively simple proteins, including pharmaceuticals such as insulin, human growth hormone, and erythropoietin. Not bad for a life-form previously known mostly for making you ill.

- 1975—Frederick Sanger develops the Sanger method for DNA sequencing, a technique that allows scientists to "read" a section of DNA.

- 1978—"Baby Louise" is the first child born through in vitro fertilization or IVF. Well into adulthood, she's still known as Baby (a nickname appropriate in a very limited number of social situations).

- 1983—Kary Mullis invents the polymerase chain reaction (PCR) protocol, which allows for rapid synthesis and copying of specific DNA sequences, allegedly with the creative assistance of a little LSD.

- 1984—Steen Willadsen performs the first successful mammal cloning of sheep from embryonic cells, and this was twelve years before we ever heard about the whole "Dolly" thing.

- 1990—The Human Genome Project is launched by the National Institutes of Health to sequence the entire human genome using a combination of cutting-edge lab work and computer science. Keep in mind, this was back when a desktop computer was significantly less powerful than an iPhone, so they had their work cut out for them.

- 1996—Ian Wilmut creates "Dolly the sheep," the first organism to be cloned from adult cells. Commence global panic in 3 . . . 2 . . . 1. . . .

- 1997—In a preemptive attempt to defeat his own clone, President Bill Clinton proposes to the U.S. Congress a five-year ban on human cloning.

- 1998—Several European nations sign a ban on human cloning, but this merely encourages the more rebellious nations to do it faster and with less caution. At the same time, the Food and Drug Administration claims authority over human cloning in the United States, implying that the government thinks a clone is either a food or a drug. We're hoping the latter.

- 2001—President George W. Bush limits federal funding of human embryonic stem cell research to cell lines that have already been created, a move that manages to limit research without adopting a clear moral stance.

- 2002—California is the first state to legalize therapeutic cloning to produce human embryonic stem cells for new medical treatments. This comes just two years after Arnold Schwarzenegger stars in *The 6th Day*, and just one year after becoming governor of California. Coincidence? You decide.

- 2003—Great Britain is the first country to issue research licenses for therapeutic human cloning. Bond has a license to kill, and now Q has a license to clone.

- 2003—The Human Genome Project is finally completed, but at three billion characters it's going to take a while to print.

- 2003—Craig Venter, founder of Celera Genomics, recreates the DNA sequence of a virus from scratch, using artificially synthesized DNA molecules.

- 2004—South Korean scientists claim to have successfully cloned and extracted stem cells from a human embryo, but an investigation in 2006 reveals the results to be fraudulent. While our capacity to fabricate humans remains limited, our ability to fabricate data knows no bounds.

- 2006—Patents are filed for *Mycoplasma laboratorium*, a bacterium whose entire genome has been artificially synthesized. The proposed organism will include a "watermark" that encodes some of the names of its creators into the organism's DNA. Can you blame 'em?

- 2007—Craig Venter publishes the first completely sequenced individual human genome—his own. Recommended summer reading. We couldn't put it down.

- 2008—The FDA approves cloned animal meat for human consumption. Now it's *exactly* like Mom used to make.

- 2008—President George W. Bush signs the Genetic Information Nondiscrimination Act, a law prohibiting the misuse of personal genetic information by employers or insurance companies. We figure he watched *Gattaca* and got *really* freaked out.

- 2009—President Barack Obama reverses the federal funding ban on stem cell research. Finally, genetic change we can believe in.
- Now—You've just purchased *How to Defeat Your Own Clone*. Or you're reading it for free in the bookstore . . . cheater.

Dystopian future:

- 2021—The first human clone is born. After her bone marrow is harvested, she is largely neglected by her "parent" and becomes a child celebrity, which predictably ruins her life.
- 2034—Clones are now genetically engineered and mass-produced. Genetic "fashions" in newborn children rapidly develop; would-be parents hold off on reproduction until Paris unveils its fall line of progeny.
- 2048—The majority of the American workforce now comprises ultraintelligent überclones. Naturally born citizens cannot compete in a job market that now requires a Ph.D. in finance to work the register at Wendy's.
- 2052—The Pentagon replaces the standing army of the United States with a single battalion of Chuck Norris clones.
- 2053—Crime and social unrest are almost completely

eradicated, but at what cost? "Norris Law" rules the
streets: every home must own a minimum of three Total
Gyms, and *Walker, Texas Ranger* plays round the clock on
all major television networks.

- 2054—Panicked, the president hastily approves Opera-
tion: Roundhouse—a legion of bioenhanced Bruce Lees.
(We can only assume they've had Lee's DNA on reserve
since 1973, stashed right between the Ark of the Cove-
nant and the Roswell alien bodies.)

- 2055—Earth lies in complete ruin after the most incredible yearlong kung fu battle ever. Don't even try to imagine
it—you'll only hurt yourself.

- 2060—Rampant ecosystem failure, exacerbated by attempts to alter our environment in the wake of the
clone war fallout, results in an Earth upon which only two
multicellular organisms can survive: cockroaches and lab
rats.

Utopian future:

- 2021—The first human clone is born. She is *ador-
able*.

- 2029—Genetically manipulated adult stem cells are now
capable of becoming any tissue in the body. Replacement
body parts can be grown on demand. Organ donor cards
are a thing of the past.

- 2034—Children are routinely born immune to HIV, diabetes, and athlete's foot.

- 2038—Through a combination of medicine, genetic engineering, and nutrition tailored to suit individual genetic profiles, people are being born healthier and smarter than ever before. Increasing life spans force people to think further than five years ahead (which is actually quite terrifying, if you've never tried it).

- 2042—Toxins in the environment are routinely cleaned up using specially tailored bacteria. L.A. finally looks (and smells) decent again.

- 2052—Modern medicine now outpaces the progression of diseases, and new ailments are cured within weeks of their discovery. The Museum of Ancient Pathology is founded to remind us what it was once like to catch a cold.

- 2053—Genetic manipulation of adult cells becomes possible. Genetic body modification quickly becomes commonplace. Pierced and tattooed grandparents worry that these kids today, with their dragonfly wings and slitted pupils, are never going to be able to find a job looking like *that*.

- 2055—Every child rides to school on a genetically engineered unicorn (trademarked "My Little Rhinopony").

- 2060—Money is grown on trees. Pigs fly.

THE FUTURE OF BIOTECHNOLOGY

No one can predict the future precisely, which is why we hedged our bets in the future part of the timeline above. Living in a Utopian future doesn't exactly require a guidebook; sipping mai tais on the beach with your genetically enhanced boyfriend or girlfriend is a no-brainer. But surviving a clone rebellion? Therein lies a challenge.

No matter what the future has in store, one thing is certain: this century's technological frontier stares back at you whenever you look into a mirror. When the atom began to reveal its secrets, there was a figurative explosion in technological progress to match the literal one in Los Alamos. Prepare yourself for the cellular version—the discovery of DNA and recent advances in molecular biology will make possible changes that are both worldwide and extremely personal. The food that you eat will be customized at the genetic level—for yield, to resist pests or pesticides, or merely for flavor, while you yourself will become the biological equivalent of a rickety bridge in desperate need of a retrofit.

The manipulation of living creatures isn't really new, of course. Farmers have been breeding plants and animals for centuries, and doctors have become expert at holding us together despite advancing age with a combination of blood pressure medication, Viagra, and hair plugs. The biotech revolution will allow humanity to design changes to biological organisms at a more fundamental level, and to implement them much more quickly and cheaply. For instance, the domestication of bananas turned the seed-filled fruits of the wild banana plant into today's sweetly flavored seedless treats found in grocery stores, but it was a long and costly process. The next generation of humans may have *hundreds* of designer fruits to top their pest-resistant breakfast cereals. With a little luck they'll be adding whole milk to that cereal, because either their bodies will be better at dealing with fats and cholesterol, or their medicine will be better equipped to deal with the consequences. Hell, might as well add a side of bacon, which like many of their pets and maybe a few of their children, will be cloned.

A few generations further on, who can tell? The future is sure to be weird and maybe even a little frightening, but we can hope that it's beautiful. If, instead, it all goes pear-shaped, at least now you've got the right book for it.

CONCLUSION: DON'T IGNORE THIS BIOTECHNOLOGY STUFF

In biotechnology there's what we call a very low "signal-to-noise" ratio as far as information is concerned. That's science talk for "you're going to need a good bullshit detector." For example, when the first cloned marmoset embryo is produced, the work will be reported by the media, whose attempt to translate technical documents and journal articles into plain English (with more than a dash of sensationalism) will edify few and confuse many. Creators of popular culture, having absorbed most of a single media headline, will immediately begin production on a film about man-eating, genetically engineered marmosets. Congress, having seen the trailer online, will then begin drafting a bill making it illegal to clone or sequence marmoset DNA, despite the protests of the marmoset-sequencing lobby.

There's a lot of misinformation out there, and we hope to prepare our readers for threats and irritations more real than imagined. As a wise Jedi puppet once said, "You must unlearn what you have learned."

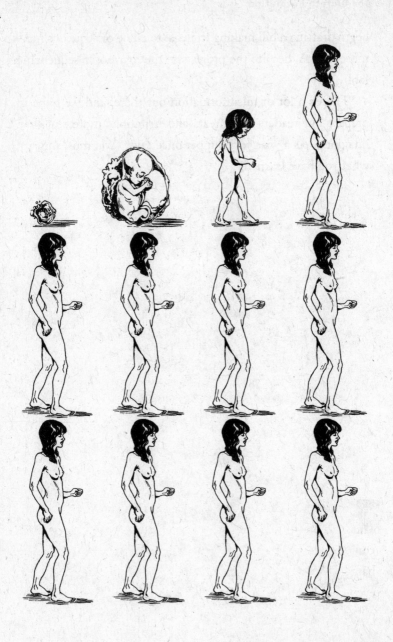

TWO

CLONING AND YOU

(and You, and You, and . . .)

*Cloning, wow. Who would have thought? There should be a list
of people who can and cannot clone themselves.*

—TED DANSON

If I was a scientist, you know what I would clone? Hot dogs!
—WILL FERRELL AS HARRY CARAY, *SATURDAY NIGHT LIVE*

The biotech revolution encompasses a wide array of social, scientific, and ethical issues, but its most intriguing and controversial aspect is undoubtedly human cloning. Highly debated and frequently misunderstood, this is a subject that everyone should be more familiar with, particularly if the "revolution" starts heading downhill.

Geneticists have come a long way since the days of Mendel

and his pea plants back in the 1800s, and the science shows no signs of slowing down. Laboratory cloning is relatively commonplace in government, academic, and industrial labs around the globe. Every day, scientists use these techniques to answer fundamental questions about life on this planet: How do organisms grow and reproduce? What are the uses and limits of stem cells? Which came first: the chicken or the egg? The answers to such burning questions may someday help us prevent cancer, spawn replacement organs in glass jars, and breed some really funky poultry.

The term clone is derived from a Greek word meaning "branch" or "twig," which refers to the horticultural process whereby a twig from a plant is used to create a new plant that is genetically identical to the original. At its core, cloning is the genetic duplication of a living or previously living entity.

If you were going to build a replica of your neighbor's house, you'd first need to understand how the original was built. Specifically, you'd need a copy of the original blueprints, plus the tools and knowledge to carry out those plans. Building a replica of your *neighbor* is no different. The process simply requires a *biological* blueprint, some very tiny tools, and a little scientific savoir faire.

DNA—A REALLY, REALLY COMPLICATED BLUEPRINT FOR MAKING PEOPLE (AND OTHER STUFF, TOO)

For any organism to develop properly, its biological blueprints must be followed precisely to ensure accurate assembly of a body with the correct number of fingers, toes, and nipples.

These instructions are stored in each of the organism's cells by a long strand of material called deoxyribonucleic acid, or DNA for short. To envision a strand of DNA, just imagine the longest chain you've ever seen; now imagine it being several million times longer. Each link in the chain is a single DNA building block called a nucleic acid, nucleotide, or simply, base. In scientific shorthand, there are four main bases, denoted as A, G, C, and T. A chain of DNA is merely an extended sequence of these bases, so a brief section of DNA might look something like this:

GGCTAGTAGCCGATTAGATTC

But how do we find any meaning in this sequence of jumbled letters? Each of us is like an enormous biological word search; there is order to the madness, but finding it demands a lot of persistence and an absurd amount of free time. Since most scientists don't want to spend their Sunday afternoons playing a billion-letter game of Boggle, we have computers do most of the grunt work. This thankless task involves reading through every nucleotide in a piece of DNA to find the "words" within, which we call genes.

A gene is simply a short arrangement of nucleotides that can be interpreted and converted by the cell into one of the many materials used for building an organism. These materials are called proteins. Setting aside a few special cases, every gene encodes for a particular protein, and every protein serves a specialized function for life.

Genes are first "transcribed" into an intermediary called ribonucleic acid, or RNA for short. If DNA is a blueprint, then RNA is a temporary photocopy of a specific piece of the blueprint that can be worked with briefly and then discarded. Imagine you need a door built in a single room of a skyscraper. Would you hand a worker the complete architectural layout to the entire building or simply give her specific instructions for the task at hand? A copy of the door schematics could be used to make a door not just once, but all over the building. When the instructions wear out or become unreadable, it's no problem—the original blueprints are waiting to be copied and handed out again. For example, a complete library of genes isn't necessary to make collagen protein; a single copy of a collagen gene will suffice. In this way, an RNA transcript is like a work order given to the rest of the cell for production of a specific protein. The actual transition from RNA to protein is performed by a complex system of molecular machines through a process called translation.

To recap, the path from genetic blueprint to biological material has two main steps: (1) DNA is transcribed into RNA; (2) RNA is translated into protein. This is how life works: if your body needs a particular material, it simply copies that section of the genetic code and starts cranking out the proteins. Seems simple enough, right? But the blueprints are massive, and the organization is nightmarishly complicated.

THE HUMAN GENOME—IT'S EVEN BIGGER AND MORE
CONFUSING THAN YOU THOUGHT

According to current estimates, humans have approximately 25,000 unique genes scattered among twenty-three different DNA chains called chromosomes. And as if that isn't elaborate enough, the genes account for only 2 percent of the total genetic material; there's additional noncoding sequences between (and even within) each gene that make up the other 98 percent. Because these extra base pairs don't actually code for any proteins, they were originally thought of as "junk DNA"— nothing more than useless filler between the meaningful information. Imagine trying to read this book if, for every coherent page of text, we also included forty-nine pages of gibberish inserted at random. Not only would the book be fifty times longer, but most of it would read like the output of one of those infinities of monkeys and typewriters trying to replicate Shakespeare.

One woman's trash is another woman's treasure, and sorting through this so-called junk DNA is like rummaging through a waste heap; somewhere between the discarded foodstuffs and soiled mattresses lie the free auto parts and vintage casual wear. The tricky part is picking out the good stuff from the garbage. This vast wasteland of DNA is littered with scads of seemingly useless information: ancient defunct genes, artifacts left by genetic parasites, and spacer sequences with potentially no function at all. But other regions have uses beyond mere protein production: noncoding RNAs that assist with various tasks, gene-control elements that regulate transcription, and many sequences that may have yet unknown functions.

The entire collection of genes in an organism, along with all of the noncoding DNA, is called a genome. In humans it's more than three-billion base pairs long. Combine that with the fact that the whole thing is smaller than a pinpoint, and you have yourself a doozy of a biological instruction manual. Think you had trouble putting that bike together last Christmas? Imagine how hard it would be if the directions were hidden throughout every book in the ancient Tibetan literature wing of the Library of Congress. And you only *sort* of understand Tibetan. And the library is smaller than the period at the end of this sentence.

GENETIC VARIATION—BECAUSE CLOTHES DON'T MAKE THE MAN; GENES DO

While you wouldn't have any trouble telling the difference between the blueprints for a skyscraper and the blueprints for a suburban home, you might have more trouble telling a pair of blueprints apart if they were both for *different* homes, because the differences between the blueprints would be more subtle— How many bedrooms? How many bathrooms? Garage or carport?

Each organism, be it a tiger, butterfly, or ficus, is the product of its species' genome: a complete set of instructions for building tigers, butterflies, or ficus. As a human being, your DNA—the instructions for building you that are shared by every cell in your body—has a lot more in common with the

DNA of other human beings than it does with the DNA of another species. Still, we don't live on a planet filled with practically identical people. Enough genetic variation exists within humanity to distinguish one person from another. At least some of this variation is due to the fact that, though you share most of your genes with the rest of humankind, your version of any one particular gene may be slightly different from your neighbor's.

Look at how humans deal with booze: every man and woman carries several genes that code for a group of proteins, which collectively help break down alcohol. For a given gene, you might have the "lightweight" version, while your friend Steve has the "fraternity legend" variety. Your gene codes for a protein that does the exact same job as the protein coded for by Steve's gene, but due to differences in its expression or subtle variations between the actual proteins produced, Steve spends his Saturday evenings doing keg stands and beer bongs with no adverse effects (long-term liver damage notwithstanding), but you end up drunk dialing your ex-girlfriend after your second strawberry daiquiri. Extrapolate this kind of variation to the other 24,999 genes in your body and you have a lot of room for diversity. The human genome makes you human; your genome makes you *you*.

To understand how this genetic variation arises, you need to know a bit about the birds and the bees, or more specifically, the eggs and the sperm (if we just lost you, please refer to the young adult section for a more, uh, "appropriate" handbook). Remember when we said that every cell in your body has the same DNA? We lied. Like any good scientific principle, there are exceptions to the rule. Eggs and sperm (called gametes or germ cells by biologists) contain only half your DNA, so that when they fuse to form an embryo, the new child receives the correct amount of genetic material: half from Mom and the other half from Dad. This is why most children have facial features that resemble a cross between their two parents (and why those that don't end up on a "Who's Your Daddy?" segment of Maury Povich's talk show).

Shouldn't siblings look exactly alike? If every child receives half of Mom's DNA and half of Dad's, shouldn't each one be an identical fifty-fifty hybrid of his parents' genomes? The answer is obviously no, but for a very specific genetic reason: every person carries two copies of each chromosome—the twenty-three bundles of DNA that make up the human genome—for a total of forty-six in each cell. That means you tend to have two copies of each gene called "alleles"—one on each chromosome pair—and they're not necessarily the same version. For example, you might have a "lightweight" allele for an alcohol-processing gene on one chromosome, but the "fraternity legend" version on the other. Your actual tolerance depends on the combination of these two gene variants.

When it comes time for the body to produce gametes, your chromosomes pair up accordingly and then separate, each new egg or sperm receiving only one chromosome from each of the twenty-three pairs. Your egg or sperm has an equally random chance of receiving the chromosome with the lightweight allele or the one with sorority lush (along with the thousands of other genes accompanying that strand). Extrapolate this arbitrary partitioning to the other twenty-three pairs, and you'll discover that a single genome can produce more than eight million unique gametes simply by shuffling chromosomes around. This means that the genetic code from two parents could theoretically produce more than seventy trillion different embryos.

Oh, and one more thing: when your chromosomes pair up, matching sections on the opposing strands can actually swap with each other, producing two entirely new chromosomes unlike anything found in the rest of your body. So that seventy trillion number was actually a bit of a lowball.

The math says there's just no freakin' way you and your older brother are going to look exactly alike, no matter how hard your parents try. They could spend the next ten thousand years gettin' busy and still not replicate that particular combination of DNA strands. On average, you will share about half of your genes with your brothers and sisters, and about a quarter of your genes with your first cousins. Your genetic relation to the rest of the world trickles off from there.

In the end, your genetic composition is governed by what your parents' gametes deal you in a game of three-billion base pair stud. So if things aren't working out in the hairline department, you know whom to talk to.

MUTATION—CHANGE IS A GOOD THING . . . SOMETIMES

While the blueprints for a house are typically unchanging, DNA strikes a delicate balance between being static and dynamic—a sort of Darwinian teeter-totter between survival of the organism and adaptation of the species. Start messing with the code and you may end up with some pretty confused cells working from nonsensical plans (our bodies need cellular co-operation, not chaos). But if nothing in the code ever changed, no new genes would ever arise and organisms would have a very limited ability to adapt.

The process by which the genetic code changes is called mutation. Potential mutations include single-base changes (like a C becoming a T), deletions (when code is lost), and insertions (when code is added). These mutations can occur through exposure to various mutagens (radiation, oxidative stress, etc.), mishaps in DNA replication when a cell divides, or even invasion by genetic parasites like viruses (which have the ability to insert their own DNA directly into your genome).

If a mutation occurs in the DNA of the cells that make up your reproductive machinery, this mutation can be passed along to your offspring. Most mutations are neutral: they don't noticeably affect the organism. Beneficial mutations give the organism an advantage, and are subsequently kept in circulation through successive rounds of breeding, so these mutations tend to stick around. Disadvantageous mutations are typically selected out of the population or are fatal to the organism right off the bat.

CLONING—THE BIOLOGICAL COPY MACHINE

In a world where perfection is unheard of and even adequacy is rare, we risk losing the occasional sublime combination of genes to the chaotic whorl of breeding. Only *cloning* can give us a precise genetic duplicate of a biological organism.

Take a drive through any modern American subdivision and

you'll notice that nearly all the homes are built from the same three or four standard tract housing designs. Voila! Architectural cloning at work. All that was needed was a copy of the original blueprints and some workers to replicate the design. But as we've already seen, it's not that simple for a complex living organism; you need a lot more than wood, concrete, and some drywall.

Imagine if a would-be cloner—we'll call him "Jimmy"—decides for some reason to make a duplicate of himself. In theory, all Jimmy needs is a copy of his own genetic blueprints and a biological device capable of building a new Jimmy from scratch. A hundred years ago, this might have sounded like science fiction, but the tools for genetic cloning are all around us; Jimmy simply needs to know where to look.

VIRUSES—NO-FRILLS ENTRY-LEVEL GENETIC DUPLICATION

Just like the plans for a house or a skyscraper, Jimmy's genetic blueprints are relatively meaningless until the instructions are carried out and the materials are organized in meaningful ways. Otherwise, all he's got is a useless pile of protein (if we only had a nickel for every time our high school gym teachers called us that . . .). To make a biological copy, Jimmy's going to need to learn how genetic duplication happens in the wild, so we'll start with the simplest version known to humankind: viral replication.

A virus is nothing more than a bundle of genetic material wrapped in a shell of proteins; the genes code for the proteins, and the proteins protect the genes. In terms of biological organization, this is as simple as it gets. But whether a virus is even "alive" is debatable. A virus is merely an infectious agent that can't reproduce on its own because it lacks the necessary machinery. Instead, a virus needs to infiltrate a more sophisticated organism and start using the available materials to make copies of itself.

A virus is a lot like an unwanted house guest. Some don't seem so bad at first, like the guy who crashes for the weekend on your pull-out sofa bed. The first night he's passed out and appears relatively harmless. But two days later he's still hanging around, and the next thing you know he's overloaded your washing machine and flooded the basement. In the virus world, these seemingly unassuming little visitors incorporate their genetic material into a host genome and may lay dormant for years before causing any noticeable problems such as AIDS. Other viruses are more like the ultimate party crasher who barges in uninvited, messes with all your stuff, and moves on when the booze dries up—except that the virus makes thousands of copies of itself and they all set fire to your house on the way out.

Irritating as they are, viruses are extremely good at one thing: creating genetic duplicates of themselves. Within the confines of a host, the viral genome is copied, a new protein shell is produced, and the parts are assembled into a new, identical virus. This process is simple only because viruses are so simple: they're outrageously tiny, use an extremely limited number of proteins, and may only carry a single piece of genetic material to be copied. Sure, it would be convenient if Jimmy could clone himself this way, but thank God he can't, or you'd probably have a thousand little Jimmys playing kickball in your yard right now. Besides, it's best not to emulate a parasite.

MICROORGANISMS—YOUR GREAT-GREAT-GREAT-GREAT-GREAT-GREAT-GREAT-GREAT-GRANDMOTHER WAS AN AMOEBA

Unlike viruses, living creatures are composed of individual structures called cells, which include a cell membrane (or wall) and a genome housed within. Sounds vaguely similar to a virus, right? But a cell is much more elaborate, capable of copying its own DNA and making its own proteins—something that those freeloading viruses can't do on their own. Like every other organism on the planet, Jimmy is constructed from individual cells, and he's going to need a way to re-create these structures if he's ever going to make his fresh new Jimmy clone.

The simplest living organisms consist of a single cell and nothing more. These bare-bones creatures were the very first recognizable forms of life on this planet, including a particular group of microorganisms that we're all intimately familiar with: bacteria. While a virus is essentially powerless without a host to exploit, a bacterial cell can transcribe its own genes and manufacture its own proteins without additional help. A bacterium also has the ability to copy its own genome and divide into two identical cells—a simple cloning process termed binary fission or asexual reproduction (which is a lot less fun than sexual reproduction, but comes in handy if you can't find a date).

Unfortunately for Jimmy, he's not a single, enormous cell, and if he wanted to clone himself like a splitting bacterium, he'd need every cell in his body to simultaneously divide and re-

organize into two identical Jimmys. As there's no known biological mechanism in place for this type of spontaneous human
cloning, this is as unlikely as it would be painful.

MULTICELLULAR ORGANISMS—
COMPLEXITY THROUGH TEAMWORK

While a single cell has the capacity to grow, adapt, and copy
itself, it takes a complex organism to paint a picture, drive a
car, or win the Stanley Cup. And frankly, this is what we really
care about, because a clone of Wayne Gretzky is going to be
a lot more interesting than a clone of the *E. coli* that gave you
the runs last night. Most of the organisms with which we're
familiar—plants and animals—comprise trillions of cells, each
with a specialized function, all working together in concert.
These cells are the fundamental building blocks of any complex
biological entity, including human beings, and they interact to
form the structures of our bodies. In order for Jimmy to copy
himself, he'll need a way to re-create his uniquely intricate
biological design, and the easiest way is to imitate the original
designer.

Mother Nature assembles organisms much as people as-
semble buildings: the genome is the blueprint for the body, and
cells are the miniature construction workers that carry out these
plans. Every cell in the body holds an identical copy of the instructions, yet each cell has a specific role in building and maintaining the complete organism. Just as a house is an assemblage
of various materials like wood, brick, steel, glass, sand, or mud,
and specific parts like walls, roof, and doors, human bodies, too,

are built from the ground up using a bunch of organic bits and pieces. In both cases, the end product is a unique structure arising from a predetermined design.

In a living system, cells and proteins come together to form tissues, such as bone, muscle, or nervous tissue. These tissues combine further to make organs: complex structures, including the heart, kidneys, or brain. And on an even larger scale, organs combine to form biological systems, like the circulatory, digestive, or respiratory systems. As a whole, Jimmy is a highly organized collection of these various systems, which coordinate to keep him alive and functional; but bear in mind that all of this complexity arises from Jimmy's individual cells and the blueprints from which they work.

If Jimmy was "cloning" his house, he'd give a copy of the original blueprints to the construction workers, and when the job was completed, the contractors would collect their paychecks and move on. Sure, they might return to do some minor repair work on the siding or to remodel the kitchen, but unless Jimmy is very wealthy, they probably won't live in his broom closet, waiting to spring into action at the first sign of a blown fuse. Cells, on the other hand, reside in an organism's tissues, where they have a number of lifelong responsibilities: they expand the tissue during growth and development, repair it when injured, make new cells to replace those lost to age and damage, and participate in the specific function of the tissue. A muscle cell contracts or relaxes to help the body to move, a liver cell pumps out enzymes that detoxify the blood, and a neuron transmits electrical signals in response to various stimuli.

In the world of the cellular "builder," every cell holds a complete copy of the genetic blueprints, but each one "reads" those plans in a different way—a phenomenon called cellular differentiation. These differences between cell types are critical, because if every possible protein in the genome was always being produced in a given cell, that cell would be a chaotic mess. Instead, each cell in the body controls which parts of the genome are turned into RNA and which are not in order to specifically control what proteins it's making. A bone cell, for example, churns out mainly bone-related RNAs and has little need for muscle- or nerve-associated RNAs. In this way, every cell type in Jimmy's body is like a specially trained worker with a very exclusive skill set—each one different, but essential to the whole—just like the builders of a house. An electrician has neither the tools nor the know-how to install Jimmy's plumbing, and a plumber would be just as baffled by the wires behind his Sheetrock, but together they can build something greater than the sum of its parts.

Let's say the construction workers are building various parts of Jimmy's clone house. They're all given a complete copy of the blueprints, but each worker requires only specific bits and pieces to do her job. The roofer needs only the top part of the design and a particular batch of materials (shingles, nails, nail gun), while another worker might need only the floor plan and tools for constructing a bathroom. Each builder needs to know not only what features are required, but how many of them need to be built: Does the bathroom need one toilet? Two? Thirty-seven thousand? This is not a trivial distinction. The same thing is true of cellular builders: each one needs to know what proteins to produce, how much of them to make, and when and where to make them in order to get the job done right. This variable gene expression is what differentiates one cell from the next, and is essential for maintaining the overall organization and function of a body. Without it, Jimmy (or his clone) might be just an enormous pile of skin cells, and that would put a serious damper on, well, pretty much everything.

The problem with trying to "rebuild" a complex multicellular organism like Jimmy based solely on his genetic blueprints is that the vast majority of cell types aren't competent to make the entire body from scratch. Jimmy can't just give a copy of his genome to a fat cell and expect it to make a new person; that cell is predisposed to making more fat. So instead of trying to recruit trillions of different cells and giving each one a copy of his DNA, what Jimmy really needs is a single cell that hasn't yet resigned itself to a particular fate. And that's where stem cells come in.

STEM CELLS—THE BODY'S JACKS-OF-ALL-TRADES

As a grown adult, most of Jimmy's cells are already fully differentiated into specific types like nerve, muscle, and bone, and for these cells, there's usually no going back. But stem cells are different—they have the ability to transform into a variety of different types depending on their current level of differentiation. The earliest and least differentiated stem cells are the embryonic stem cells—found in the embryo—which can theoretically become *any* cell type in the body. They can also divide into two identical stem cells, each with the ability to either divide again (like the "broomstick" scene from *Fantasia*), or differentiate toward a particular lineage.

When an egg and sperm first combine, the result is the ultimate stem cell: a zygote. All the other cells in an organism are derived from this one original cell. The steps from "single cell" to "baby Jimmy" could (and do) fill a catalog of textbooks, but the general idea is simple: the zygote first divides exponentially to make a cluster of embryonic stem cells. Some of these cells then begin to change and differentiate to form the early parts of the Jimmy fetus, while others continue to divide and replenish the stock. As the process continues, cell division allows the developing Jimmy to grow, and cell differentiation helps him acquire his Jimmy-ish shape. After nine months, a single Jimmy zygote has grown into a complete infant Jimmy—two arms, two legs, two eyes, and a boatload of necessary organs. The point is that complex organisms don't exist as a single cell, but they all *start* as one—a very special one—and this phenomenon is what will allow us to replicate the development of a specific individual.

NUCLEAR TRANSFER—A CLONE IS BORN

If a lone zygote can make a complete Jimmy from a single copy of his genome, then why not use that mechanism to make a *new* Jimmy (or several new Jimmys)? In nature, an embryo is formed through the chance meeting of a unique egg and sperm, the result of which is a zygote with a never-before-seen and never-to-be-remade genome. But the genome still exists for as long as that organism's DNA remains intact, and re-creating a zygote is *much* easier than re-creating an entire person.

The technique for making a "copied zygote" is called somatic cell nuclear transfer, in which the nucleus (the part of the cell that holds all the DNA) of an unfertilized egg is removed and replaced with the nucleus from a somatic cell (i.e., any adult cell other than an egg or a sperm) from the organism to be cloned. To create Dolly, scientists took the nucleus from an adult sheep's mammary gland cell and transferred it into an enucleated egg from a Scottish Blackface ewe. The result was a single egg cell containing a complete set of genetic blueprints for making a copy of the original sheep from which the mammary cell had been taken. That cell was inserted into the womb of the ewe, where it developed naturally into the clone we knew as Dolly.

To clone himself, Jimmy would have to re-create this nuclear transfer process for a human: he would need an egg cell from a female donor (any healthy egg will do) and a nucleus from one of his own cells. After removing the nucleus from the egg and inserting the DNA from his own cell, the egg now has the genetic material that Jimmy's original zygote had and is tricked into thinking: it's time to make a new baby Jimmy. All Jimmy needs now is a surrogate mother willing to carry around his developing clone for the next nine months. (Good luck, Jimmy.) Ladies, at least you have the option to be your own clone's surrogate mother, although this will probably make defeating her even harder, what with all the emotional attachment.

While it's essentially impossible to precisely copy several

trillion cells and piece them back together into something Jimmy-esque, it's far less difficult to make a single cell that's primed to remake Jimmy and let Mother Nature do the rest of the work. At the moment, there are still several technical, ethical, and regulatory hurdles to overcome in terms of *human* cloning, but the groundwork has already been laid in other species, which means that it's only a matter of time before Jimmy starts cranking out copies of himself into an unsuspecting world.

You may have a wiser head than Jimmy, and you may be inclined to think twice before cloning yourself, but that doesn't mean someone *else* can't clone you. Your body is the unique source of your genetic blueprints, but it's constantly shedding copies—skin flakes, a drop of blood, or even a single hair follicle can be enough to provide a complete, clonable genome. Anyone with a basic knowledge of molecular biology and the right lab equipment could theoretically obtain a copy of your genomic DNA. Once human cloning technology becomes a reality, these DNA thieves will need only a donor egg and a surrogate mother to release a copy of yourself onto an unsuspecting you.

NATURAL HUMAN CLONING—MOTHER NATURE IS NOT IMPRESSED; SHE'S BEEN DOING THIS FOR AGES

Many of our readers probably already know this, but for those who don't, let us set the record straight: identical twins are clones. Occasionally, when a fertilized egg divides to make more cells, those cells may physically separate, splitting into two or more identical embryos in the womb. Because these embryos share the same genetic code and because the cells are still in the "embryonic stem cell" phase, the duplicated embryos start building from the same set of genetic blueprints, and thus identical twins are born: genetic duplicates traveling along parallel paths of development.

The neat thing about cloning is that it's essentially like twinning, but with the two twins potentially living on disjointed timelines. Before Dolly the sheep and the advent of reproductive cloning, if you wanted to clone yourself you'd have to make that decision as an embryo, which is quite a big decision to make during your first trimester. In the future, you'll be able to make that decision (foolish as it may be) whenever you want. Just remember that unless additional technologies are created to artificially age your clone, your genetic duplicate is going to start life as an infant no matter how old you currently happen to be.

HOW TO SPOT A CLONE BY USING SCIENCE

Imagine your best friend is in a fight for her life against an evil clone version of herself. And there you are, holding a gun with just one bullet left. Whom do you shoot? What was once a dilemma only for militant, heavily armed acquaintances of identical twins, will someday, as a result of cloning, become slightly less incredibly rare.

One can identify a clone by looking for the same differences that a mother uses to discriminate between her two genetically identical twin daughters. Many of the unique traits that she uses to distinguish her offspring arise from differences in their development, and are just as likely to appear in clones.

Genetic anomalies

Random localized cell mutations (those starting in a single cell) during development, or even later in life, can result in skin variations such as birthmarks, moles, or freckles. Identical twins rarely share the exact same set of "skin tags," so it's unlikely that your clone would share yours.

Mutations that occur during the early stages of development are present in every cell in the body. In a few reported instances, one twin is perfectly healthy while the other is born with Down syndrome or some other biological alteration. A similar genetic glitch during cloning could wreak havoc on your would-be replica.

Developmental variations

Between the time your parents got frisky and the moment the doctor slapped you on the behind, you spent about nine months doing . . . what exactly? Well, you were feeding, growing, and otherwise managing one hell of a metamorphosis from a single cell, all the while being a royal pain in the uterus for dear old Mum. Needless to say, the time spent in the womb is critically important, because even the smallest change in this prenatal environment can have a major impact on those embryonic cells. Smoking, drinking, diet, exercise, and the overall health of the mother during pregnancy can all have an effect, ranging anywhere from birth defects or miscarriage on the negative side, to increased intelligence or physical fitness on the positive. Differences between your clone's prenatal environment and your own—either through use of a surrogate mother or an artificial womb—may result in subtle (or not so subtle) differences between you and your genetic double.

If you need a more concrete example of the importance of prenatal environment, look no further than the sci-fi cartoon masterpiece *Futurama*. When Professor Hubert J. Farnsworth created a young clone of himself (Cubert Farnsworth), the duplicate's appearance differed from the professor's because his nose was squashed up against the wall of his cloning tube during development. Just one more reason to stop growing things in tubes. We need spheres, people. Spheres!

But even if you could get the same womb for your new double ("Hi, Ma, I've got a small favor to ask . . ."), it wouldn't necessarily yield identical results. Your specific prenatal environment was a unique function of time and space, and it won't be easy to replicate. If Mom was twenty-three and healthy when she was carrying you, but forty-seven and hitting the bottle when you were renting her womb for your clone, you'd be looking at some serious quality-control issues.

Even identical twins developing in Mom at the same time don't necessarily experience the *exact* same prenatal environment. If the twins separate into two embryos early in the pregnancy, they each form their own chorion (outer membrane) and amnion (inner membrane), effectively dividing the womb into two separate environments. It's the prenatal equivalent of splitting your dorm room down the middle with tape to make sure that your roomie stays out of "your side." Sure, you both live in the same apartment, but his side may have better access to the bathroom while yours is closer to the minifridge. Despite your proximity, your environment and your roomie's are different. Twins that separate later in pregnancy can share a chorion yet maintain their own amnion, or later still, share both chorion and amnion. These sorts of differences in the confines of a single womb can influence how identical twins develop early on, and further demonstrate the influence of the prenatal environment on human development.

Without a great deal of work to precisely reconstruct your entire development, your clone can—and likely will—be substantially different from you in many ways.

Epigenetic changes

While a mother's womb provides the primary setting for early development, nine months is a mere drop in the tank compared with a lifetime in the outside world. Even though our personal library of genes is essentially fixed at conception, the specific way in which these genes are expressed (where, when, how much, etc.) is mutable.

In biology, alterations in gene expression that occur without changing the underlying code are called epigenetic. Some of these modifications can be inherited from your parents, but others happen later in life and are unique to you alone. It's already clear that epigenetic changes can have an impact on your metabolism, and if your metabolism is significantly different from your clone's, chances are your body shape will differ as well.

Donor-cell disparities

Identical twins are produced when a single fertilized egg splits into two duplicate embryos, but a lab-made clone is slightly different: The individual to be cloned must donate DNA from one of his or her adult cells in order to create a new embryo. However, your DNA continually changes throughout your life, mutations accrue, and genes are epigenetically silenced. These and other alterations make the genetic condition of each of your adult cells slightly different from their original embryonic form.

The degree to which your DNA has changed depends on the age and state of the cell from which it's taken. Imagine removing the engine from a twenty-year-old 'Vette and dropping it into a new frame. Is this new car a "clone" of the original? Sort of, but it carries with it some of the unseen wear and tear of its older counterpart. Analogous "depreciation" in the nucleus of your donor cell may result in a human clone that's not quite a perfect replica of the original. Especially because DNA mutations can accumulate over time, it may become increasingly difficult to find a suitable donor cell as you get older.

Even if you can find a donor cell with immaculate DNA, that DNA is only half the equation; to clone yourself, your DNA needs to be fused with an egg cell. This egg contains "mitochondrial DNA" that exists apart from the nucleus—an exception to the standard rule of the genome, and something that is passed directly from mother to child during normal human reproduction (tough luck, guys, but you've still got the Y chromosome, so don't let it get you down). The extra code from this mitochondrial DNA will be different for your clone unless you can get your mom in on the deal (and trust us, if she's anything like our moms, she's definitely *not* in). So, even if you manage to get pristine DNA from a donor cell nucleus, the egg "chassis" you're dropping that nuclear engine into may be different from the one you rolled out of the factory with.

Random developmental events

Although scientists don't often admit it, there are lots of things we can't predict, especially when it comes to human development. The genes in your DNA dictate what proteins can be made, but not necessarily when to make them, or where they should go once they're hanging around. Getting a gene to make a protein requires a complicated series of chemical reactions to occur, involving dozens of different molecules, many of which are present in only relatively small amounts. One cell might complete the process in record time while another lags behind, due only to random variations during the chain of events.

The scientific term for something that displays this sort of inherent randomness is stochastic—this comes from a Greek word meaning "guess," and guessing is basically all you can do. Consider the formation of a snowflake. Each individual water molecule is obeying the same laws of physics as all the others. As the crystal begins to form, random, seemingly insignificant variations start to appear. One crystal packs on a few water molecules in one configuration while another crystal adds the molecules in a different pattern, giving the two flakes slightly different shapes. These distinct formations affect how subsequent water molecules attach. All of these random variations accumulate, building upon one another, until each flake is visibly different from the others.

Fingerprints are the body's snowflakes. Even though identical twins share the exact same genetic blueprints, they don't share the same distinguishing pattern of grooves and ridges on the tips of their fingers, because there's no explicit set of genes for fingerprints. Rather, there are genes that produce specific structural proteins for making skin, and there are other genes that make signaling proteins for determining "a fingertip goes here." The rest of the puzzle is completed through a series of complex interactions involving natural laws and random chance. It's the "random chance" part that makes things tricky and really confounds our attempts to guess everything in advance. Interestingly, twins' fingerprints are different enough to tell apart, but are still more similar to each other than nonidentical brothers and sisters or unrelated individuals.

Random variations arising during development are not limited to fingerprints alone. As eyes develop, a network of blood vessels forms to keep the retinal cells healthy and nourished, though the microscopic shape of this network is not laid out in advance. The formation of the blood vessels is not completely uncontrolled, but there's a lot of leeway—the body doesn't need to micromanage the blood vessels to properly feed the cells. Just give the blood vessel–building cells a few general rules: branch out occasionally, avoid putting blood vessels where we don't want them, and stop when the network is adequate to keep the cells happy.

To identify a clone, a twin, or any other person for that matter, you could use a retinal scanner to examine the specific configuration of blood vessels on the retina or an iris scanner to check the unique pattern of ridges on the colored part of the eye—features which depend more on random developmental variation than on genes. And, of course, fingerprinting technology has been around for decades.

Unfortunately, you'll need some sophisticated lab equipment to identify these details in a genetic double, but it might be worth it if you're in a pinch. Our recommendation is to put that retinal scan on file now before your clone beats you to it. In the end, your genome can be copied, but the precise series of cellular events that built you cannot, and that just might be enough to help you spot a rogue clone.

NATURE VERSUS NURTURE—
BIOLOGY'S HEAVYWEIGHT TITLE BOUT

You probably don't live in a bubble (unless you do), so genes are only part of your developmental story. The rest depends on external influences from your surroundings that help shape your body and personality. You will inevitably make changes to your body (both reversible and irreversible) that cannot be duplicated by genetics alone. The absence of the following alterations is another great way to distinguish a clone from its counterpart:

- Injuries—Scars, broken bones, crooked joints, rheumatic knees, even amputated limbs, are all unique physical characteristics. Remember that time you fell down the stairs and ended up with sixteen stitches in your forehead? Your clone doesn't. And unless she's willing to relive all your past calamities, you've got some pretty wicked distinguishing marks without even realizing or working at it.

- Tattoos—These are easier to replicate than scars, but still take time and effort. And what about that tattoo you've been regretting? Maybe your clone will think twice before getting your ex-girlfriend's name tattooed on his ass . . . even if you didn't.

- Dental history—When all else fails, dental records are the easiest way to identify a corpse, and the same may be true for your clone. Fillings, bridges, tooth loss, and braces all add up to your current dental condition. Even the color and the wear on your teeth reflect years of diet and maintenance. You might have thirty years of coffee stains, but your clone comes with a fresh set of pearly whites to discolor as she sees fit.

- Surgeries—While some surgeries leave only a small scar, others radically alter your appearance. Breast implants, rhinoplasty, face-lifts, liposuction, and other cosmetic procedures create changes beyond the limits of your genes. Even concealed surgeries such as an organ transplant or a biomedical implant can be used as lasting markers to differentiate you from your clone.

- Hair—Any accomplished fugitive will tell you that the length, color, and style of your hair can be easily changed to modify your appearance. Is your hair short or long? Dyed, bleached, or untreated? Styled or unkempt? Facial hair, too, can drastically change the way you look. If your clone doesn't have time to braid those dreadlocks or grow out that handlebar mustache, it should be fairly easy to tell you two apart.

- Skin pigmentation—A simple tan or sprayed-on color can distinguish you from your clone. Just don't overdo it or you'll be needing that "resistance to radiation" gene before it's ready (see Chapter 4: Bioenhancements).

- Diseases—Many illnesses leave a mark even after they've run their course. Take chicken pox for example; you've probably been immune since childhood, but it's not something that's encoded in your DNA. This resistance is a product of your immune system's "memory" to the diseases that you've already encountered. There are also those diseases that stick around for a while, or, if you're unlucky, forever. A simple blood test could determine who's the original and who's the clone.

- Fitness level—Could Arnold Schwarzenegger's clone pass for the Terminator if the clone spent his life on the couch scarfing Mallomars instead of working out? Unlikely. If you've led a healthy life, then your clone will probably need some exercise before he can fool anyone. On the other hand, if you've let yourself go, then he'd better start eating.

Perhaps even more than your body, your mind is a blank slate at birth. A clone of Einstein *might* be predisposed toward math and physics, and a clone of van Gogh *might* gravitate to art and self-mutilation, but given different circumstances they might both grow up to be freelance music critics. When it comes to personal development, everything's thrown into the mix and any single factor may be enough to tip your clone's hand. Your education, where you were raised, every experience in your life, no matter how banal or traumatic, become part of who you are—and who and what your clone isn't.

If you separate a pair of twins at birth, raising one as a soldier and the other as a baker, you will not end up with two hardened warriors each capable of whipping up a mean croissant. More likely, you'll get a commando and a pastry chef with eerily similar passport photos. You and your clone will tread the line between intrinsic genetic programming and external influences. Your clone may share your taste for redheads, but she won't share your first kiss. She might possess your innate hand-eye coordination, but she won't know how to play Xbox like you do (though she may be just as much of a sore loser).

CONCLUSION: KNOWING IS HALF THE BATTLE

Let's say you need to demolish a building; you could nuke the hell out of it, or you could take advantage of your architectural and engineering know-how to discreetly target its main support pillars. When facing the dangers of the biotech revolution, your actions should be no less calculated. A general knowledge of biotechnology will give you a serious strategic advantage.

THREE

COMMON MISCONCEPTIONS ABOUT CLONING AND BIOTECHNOLOGY
(Popular Culture Is a Poor Teacher)

Anybody who objects to cloning on principle has to answer to all the identical twins in the world who might be insulted by the thought that there is something offensive about their very existence.
—RICHARD DAWKINS

The cloning of humans is on most of the lists of things to worry about from Science, along with behaviour control, genetic engineering, transplanted heads, computer poetry and the unrestrained growth of plastic flowers.
—LEWIS THOMAS

ow that you know what a clone is, it's time to go over what a clone *isn't*. Popular culture has been misrepresenting clones since the term was applied to *Homo sapiens*. If you want to make the most of the biotech revolution, you're going to have to unlearn the most egregious of culture's misapprehensions.

A CLONE WILL KNOW EVERYTHING THAT YOU KNOW, AND WILL ACT AS YOU DO

Clones are often depicted in popular culture as experiential as well as genetic copies, sharing some or all of their counterparts' memories, even if the originals are long deceased. In the final (and lamentable) installment of the *Alien* tetralogy, clones of an alien-Ripley hybrid have access to the memories of the original Ellen Ripley. In *Aeon Flux*, the members of a society that has for many generations been composed entirely of clones possess shadowy memories of their past "lives." And in *The Fifth Element*, Milla Jovovich's character is almost completely vaporized, save for her right hand; lucky for Milla, the genetic engineers of the future need little more than a few strands of DNA to completely reconstruct her body *and* mind. In reality though, this technique would generate at best a genetic duplicate of Milla's body. Her thoughts and memories are contained within the neural connections in her brain, formed during a lifetime of sensory perception: such experiences are not stored in an individual's genetic code. Even if Milla's dismembered hand survived, any memories of her sweet-sixteen birthday party—and everything else for that matter—would be long gone. The biotech revolution promises a number of immediate identity crises, but a genetic clone that also shares your mind is not one of them . . . yet.

Granted Milla's character was an alien with supposedly "perfect" DNA, so perhaps we should give her the benefit of the doubt. You, however, are human, and unless you're storing backups of your memories in your phalanges, cloning won't bring you back from the Great Thereafter, let alone create a clone that remembers all your email passwords.

Still, what if you *could* copy your mind as well as your body? Surely this would allow you to come back from the dead, right? In *The 6th Day*, Arnold Schwarzenegger repeatedly slays a set of unrelenting pursuers, only to find them replaced by seemingly identical enemies after every kill. The villains accomplish this perceived immortality through a pseudoscientific process of copying the memories from their own recently deceased brains and introducing them into the bodies of their brand-new clones. Alas, science has yet to imagine a way to transfer memories from a corpse into a clone. Even if the technology existed, would this new clone really *be* you, or merely a copy who *thinks* he's you?

CLONES ARE DECANTED AT THE SAME AGE AS THE ORIGINAL

If you want a clone similar in age to yourself for a long-lost twin, emergency kidney donor, or alibi generator, be prepared for a dip in liquid nitrogen. Your best bet is to cryogenically freeze yourself, hoping that the technology for a successful thaw has been developed by the time your clone catches up with you. Proponents of Einstein may argue that it's easier to flirt with the space-time continuum if you want to temporarily suspend your age, but since none of these proponents has a ride badass enough to travel near the speed of light, the point is moot. Even if you manage to give your clone a chance to catch up with you in the age department, the danger doesn't end there. Your clone will have to be raised by someone whom you trust implicitly, and there's always a chance that she'll simply leave you in the tank and take over your life.

Accelerating a clone's development is technologically far more challenging than postponing your own. In theory, scientists might be able to harness certain diseases that accelerate aging, such as progeria or Werner syndrome, but the progression of aging is dangerously abnormal (victims of these diseases appear much older than their true age, and typically die early in life). The rate at which we develop runs deep in our nature, and even if we're capable of aging the clone in utero so that he is the same age as you (which is a mighty big step), we still have to do something about his learning (which is a mighty bigger step). Unless you can teach the clone in the tank at a comparably expedited rate, you'll end up with an adult duplicate that drools and messes himself, and there's nothing more embarrassing or unsanitary than an attempt to take over the world with an army of clones that missed out on potty training.

Keep in mind that even if we could automatically teach a clone the basics of life—how to feed himself, tie his shoes, and program a TiVo—the basics are not what makes you *you*. Knowing how to tie your shoes doesn't give you the "muscle memory" of a lifetime avoidance of Velcro trainers. Learning a language from a textbook does not make you a fluent or engaging speaker.

A CLONE HAS NO SOUL

The authors have no desire to delve into a centuries-old metaphysical debate. But we will say this: a clone will have just as much of a soul as an identical twin or a baby born from in vitro fertilization. No more, no less.

CLONES SHARE A TELEPATHIC LINK WITH EACH OTHER

Unless you're already telepathic, this is not very likely. Chances are you will not know when your clone is in danger or feel your clone's pain, and for that we ought to be thankful. Defeating a telepathic clone is strictly a graduate-level project. Let's learn to crawl before we walk.

CLONES ARE CONVENIENT ORGAN FARMS

On the contrary, clones are both expensive and potentially dangerous when mistreated. While it's true that your clone would share your genetic code, thereby circumventing the problem of organ rejection, there are several pitfalls to consider:

- Your clone is both craftier and more elusive than a heart or a lung. On the day of your surgery, a lung cannot hide; your clone can and will. Plus, catastrophic multi-organ failure is rare, so you'll probably need only one spare organ from your clone at a time. Keeping an entire clone around for your very occasional organ needs is not cost-effective. It's also more than a little repugnant. Those who are known as clone-killing organ-filchers do not win friends or influence people.

- Unless you keep your clone in a vegetative state, she is likely to oppose the whole "organ donation" idea. Can we keep our clones comatose over the course of a lifetime? Maybe. But consider the cost of renting a bed in a nursing home for the next half century. Even if money is no object, you're still looking at using an organ that's been wasting away on old linen for decades. In this light, taking your chances with the kidneys you've got doesn't look so bad.

- Infant organs are going to be of no use to you (and for that perhaps we can all be a little grateful). Are you willing to plan your every medical need two decades in advance? No? We thought not.

- Anyone who has seen *The Island* or *Parts: The Clonus Horror* knows that it is virtually impossible to keep a clone ignorant for her entire lifetime. That's a feat that can be accomplished only by a steady diet of reality television.

To sum up, cloning for organ embezzlement is one of those situations where you're better off not being a bastard. By the time your clone's organs are ready for transplant, the medical establishment is likely to have a legal and ethical alternative to stealing her organs that will work just as well (or better). Bottom line: look into personal genetic enhancement instead, and support research into tissue-engineered organs.

A CLONE OF YOURSELF OF THE OPPOSITE SEX
WILL BE TOTALLY HOT

Sadly, this is not the case. Preliminary results from our exper—er . . . our *friends'* experiments indicate that your morphology will be made no more physically attractive by swapping a Y chromosome for an X or vice versa. All trials thus far demonstrate statistically insignificant levels of increased badonkadonk.

On the off chance that a gender-reversed clone of yourself is particularly attractive, remember, dude, that's effectively your sibling. If awkward is ever an Olympic event, you're looking at a silver medal at least.

CLONES ARE INHERENTLY EVIL

If you think that your clone—a being who shares your DNA—is destined to turn evil, all we can say is: Have you considered therapy? A clone raised in an environment similar to yours is likely to be approximately as evil as you are; a clone that is raised in a government-run orphanage for assassins may have more of a tendency toward violent mischief.

THE XEROX EFFECT

In the film *Multiplicity*, Michael Keaton's clone clones himself to create a clone twice removed from the original. This second-generation replica suffers from a severely diminished IQ, which is rationalized by stating that "when you make a copy of a copy, it's not as sharp as the original." But cloning is a lot more complicated than making a Xerox or dubbing a mix tape. During development, your unique genetic code is replicated trillions of times under a tightly coordinated system of checkpoints and fail-safes. Any cell failing to reproduce its DNA precisely is likely to be terminated before it can cause trouble, leaving you with cells that all share the exact same genetic code (with a few special exceptions). And much of the so-called "junk DNA" can be mutated with relatively little effect. Likewise, the genetic code itself is redundant: some changes—even in bits that are expressed as proteins—won't change the sequence of the resulting product. So even a genome with a few bloopers might still result in a functional clone.

While most mutations are effectively neutral, over a lifetime some of our cells may accumulate detrimental changes that slip by the cell's error-checking machinery. This is why it wouldn't be a great idea to make a clone using DNA from your rectal tumor. Its genetic material will be a complete mess . . . and gross. Note, however, that none of this damage is specific to cloning; so while a botched clone can happen, there's no good reason to think that successive clonings would make it more likely. Besides, if your clone is off somewhere making clones of herself, you have bigger things to worry about than whether or not they possess the charm and wit of the original.

YOUR CLONE IS MORE AFRAID OF YOU THAN YOU ARE OF HER

This might be true for snakes, spiders, and small rodents, but clones? Let's not get carried away.

A CLONE HAS NO BELLY BUTTON

Was your clone created outside a womb? If not, he'll emerge with his umbilical cord. Even if your clone does spend his first nine months in a glass jar, the attending doctors will probably be clever enough to fake a belly button. But this vestigial structure is nothing more than a scar, so its appearance has very little to do with genetics; identical twins can often be distinguished by differences in their navels. Your clone won't necessarily lack a belly button, but he probably won't share the same one as you.

"CLONING" IS THE SAME THING AS "DUPLICATION"

Imagine this: A machine is created that can precisely copy the exact physical makeup of any object, atom for atom. You step inside the machine, flip the switch, and a flawless replica of yourself pops out the other side. The new copy is a perfect duplicate in every way, but this is not genetic cloning in the traditional biological sense.

Cloning requires only the replication of your existing genetic code, not of your precise atomic makeup. The hypothetical duplicating machine could easily make a copy of an inanimate object like a brick or a tube sock, but biological cloning of these items would be impossible for even the most gifted geneticist. Plants, animals, and anything with a double helix are amenable to cloning; motorcycles and twenty-dollar bills are, unfortunately, not.

On the other hand, the process of genetic cloning is easy

when compared with atomic duplication. Nature has been kind enough to provide us with all the intricate machinery we'll ever need to replicate DNA: a single cell. Sure, it took us a few thousand years to figure out how to use the replicating contraption known as the female egg, and it certainly isn't perfect (three trimesters of pregnancy plus labor is a long time to wait for your new double), but it's better than anything else we've come up with so far. As for your atomic duplicator? Alas, the world's physicists and engineers have been wasting time working on things like solar power and the computer, leaving us bereft of such a device.

Just remember: while any duplicate would have to be a clone, not every clone is an exact duplicate.

IF TWO CLONES COME INTO PHYSICAL CONTACT WITH EACH OTHER, THEY WILL IMPLODE INTO NOTHINGNESS

Let's try to stay focused here. This is a book about cloning, not alternate universe antimatter twins. If you want a book about the physics of *Timecop*, the authors suggest you find an exceptionally creative physicist with basic cable and a lot of time to kill.

CLONING WILL DECREASE BIODIVERSITY AND CAUSE OVERPOPULATION

Biodiversity can refer to either genetic diversity within a single species or species diversity within an ecosystem. A lack of biodiversity is a real problem, leaving a population at risk from disease or disaster, but it's not a problem unique to cloning;

large-scale farming already does much the same thing by breeding stocks for advantageous attributes. In fact, cloning and related technologies can help maintain biodiversity by keeping endangered species alive or by preserving rare genes. A genetically diverse human population is likewise advantageous, though it's extremely unlikely that cloning will become so prevalent that it will substantially reduce human genetic diversity. The gene pool has thus far tolerated the occasional twin or triplet, and unless cloning becomes a common alternative to sexual reproduction, it's unlikely that a few clones here and there are going to ravage our genetic diversity or lead to overpopulation. And let's be honest—sex has a lot going for it.

IF LEGALIZED, HUMAN CLONING WILL ALLOW GOVERNMENTS TO CREATE AN ARMY OF GENETICALLY SUPERIOR CLONED WARRIORS

Hmm, this depends . . . Which government are we talking about? Switzerland or North Korea?

IT WOULD TOTALLY SUCK TO BE A CLONE

Switch roles for a second and imagine yourself as the clone. With all the negative attitudes in the public eye, there's a good chance you'll be treated like a second-class citizen, existing only as a "fake" person or just a "copy of someone else." Or, like the first test-tube baby, by the time you've grown up no one will think it's a big deal anymore. Being the clone of an authoritarian egoist or the president's body double would probably suck;

being Will Smith's identical stunt double could be pretty cool. Clones will simply have to make do with the chromosomes and upbringing they've been given . . . just like the rest of us.

Of course, if you and your clone are constantly trying to defeat each other, it'll suck to be either one of you. Can't we all just get along?

THE ONLY WAY TO GET STEM CELLS IS FROM ABORTED FETUSES

One popular alternative to fetal embryonic stem cells is adult stem cells, which can often be harmlessly extracted from living adult donors. Unfortunately, their potential is limited; a neural stem cell might be used to produce new neurons or spinal-cord tissue, but it won't help you fix a damaged spleen.

The allure of embryonic stem cells is that they can become *any* cell type. Theoretically, you could take a single embryonic stem cell and use it to reconstruct any organ in your body. Practically, an adult stem cell can be much easier to work with, with one added bonus: you don't have to explain to a patient that her next arthritis treatment is fresh from a family-planning clinic. To many patients that's basically the equivalent to saying, "I have a cure for your heart disease. Just eat this bag of puppies."

STEM CELLS CAN CURE ALL DISEASES

The great promise of stem cells is that they have the potential to generate new tissues, organs, and cells that could be used to replace damaged, aging, or defective ones. Diseases or traumas (such as Alzheimer's, deep tissue burns, and spinal-cord injuries) that cause harm to the body's otherwise healthy tissues, could theoretically be treated with stem cells. In fact, hematopoietic (blood making) stem cells have already been used for decades to treat leukemia and lymphoma through a procedure known as a bone marrow transplant.

Unfortunately, stem cells aren't natural infection killers. They're not designed to eradicate the AIDS virus or even clear up that pesky swimmer's ear. To do that, we need antivirals, antibiotics, or new emerging treatments to rid the body of these unwanted maladies. A healthy stem cell might be able to replace an infected cell, but you still need a way to clear that infection in the long run. This being said, stem cells might be used to replenish the cells of a weakened or deficient immune system. Still, that's not necessarily a cure. In other cases, stem cell research may give us new insights into the inner workings of developmental diseases (cancer, birth defects, etc.), but the cells themselves might not actually be used in the ensuing treatment.

Sure, it would be great if stem cells could cure your athlete's foot. It would also be nice if they gave you a little foot massage while they're down there, but it probably isn't going to happen.

DECAPITATED HEADS CAN BE KEPT ALIVE INDEFINITELY

Historically, there has been a rather morbid fascination with post-decapitation consciousness. Back when the guillotine was a fusion of capital punishment and public entertainment, various "experiments" upon recently severed heads were devised to determine how long a head remained aware when detached from the body. Our favorite: slapping the severed head to see if it gave you a dirty look. (Seriously, though, that is not cool.)

While it's humbling to admit it, much of the body is little more than a vessel for transporting the brain from place to place and keeping it supplied with oxygen and nutrients. The mind-body connection is a complicated one, but what most people call their "self" is defined more by what's going on in their head than anything else. As long as the brain has a means to survive, one could theoretically preserve consciousness and extend life beyond the potential limits of the body.

There's a video floating around on the Internet of an experiment from the 1940s in which Russian scientists are shown using an intricate system of tubes and pumps to keep a dog's severed head alive and conscious for several minutes (bear in mind, this was back when things like "scientific regulations" and "codes of ethics" were just *words*, man). Whether or not the video is genuine is debatable. Even if this technique was real, that's a lot of trouble to go to for a few extra bodiless minutes of existence.

Variations on this theme have existed for decades. Many of *Futurama*'s contemporary guest stars are alive in the far future thanks to head-in-a-jar technology. But how exactly would one replicate this crazy scheme? Even if you could formulate the perfect artificial solution for eternal brain health, the human cranium isn't designed to be submerged in oxygenated sugar water. The ions and pH levels alone would probably be enough to eat through your skin in a matter of weeks.

Conclusion: decapitated heads don't survive very long and, even if they did, "an inability to scratch one's nose" and "gibbering insanity" would top our list of reasons not to do this.

BIOENGINEERED SHARKS—
SMART AS HUMANS, EVIL AS SHARKS

It seems like every time movie stars try to save (or take over) the world with bioengineering, it always ends in catastrophe. Take the film *Deep Blue Sea* for example, in which a team of researchers genetically enlarges the brains of three mako sharks in order to harvest a particular enzyme for use in Alzheimer's treatment. Well, the new genius sharks don't take too kindly to having nuggets of their brains removed, so they outsmart and devour the scientists instead.

But why did they have to become genius *killer* sharks? Why not a league of sophisticated gentlemen sharks sipping highballs and discussing the Dow Jones Industrial Average? Well, for one, it doesn't make for very engaging cinema. For another, the scientists didn't use much foresight. They could have just as easily created sharks with huge brains and no teeth. The consequences of our genetic tinkering are all in the details, folks. If we're concerned about the bioenhanced creatures of the future, we don't need *less* engineering, just less *stupid* engineering.

THERE IS A SINGLE GENE FOR EVERY TRAIT

You're probably already aware that the X-Men have a "mutant gene" that gives them their superpowers. Genetic intricacies aside, common sense should tell us that this is somewhat preposterous; if it's a single gene, then why do they all have completely *different* powers? That would be like saying there's a single gene for the human face—one that has six billion different variants. A more reasonable (and more accurate) explanation is that various combinations of genes work together to make complex structures and features. Granted, sometimes very simple traits like a dimpled chin or the ability to roll your tongue are controlled by a single gene, but most things in the human body are far too complicated for that.

So the next time someone tells you that there is no gene for the human spirit, you can politely agree. But you can also inform her that there might be a vast collection of genes that, when working together, create a neurological condition generally approximating something that we commonly perceive to be the human spirit.

Just remember: there is no single gene for the human spirit, but neither is there a gene for the human armpit. Or the human nostril. Or the human ability to embarrass oneself at an office Christmas party.

GENETIC SCREENING CAN FORESEE YOUR MEDICAL FUTURE

There's a lot of talk these days about the socio-ethical dilemmas posed by genetic screening, and for good reason. The general consensus is that a comprehensive survey of your genomic flaws and shortcomings will be a mixed blessing, with your future health problems becoming available to potential colleagues, employers, and medical providers alike, long before any of those problems occur. On the one hand, this could give you time to mentally prepare yourself for your future membership in the Hair Club for Men. On the other hand, it could leave you with obscenely high medical premiums and the disquieting details of your own genetically determined death. While the whole insurance-companies-are-dillweeds thing is probably true, their decisions will be based on probabilities and statistics, not certainties, because genetic predictions are just that: predictions.

In discussing the previous misconception, we explained how genes often work in combination to make complex traits. The technical term for this sort of genetic "teamwork" is polygenic inheritance, wherein a particular trait—autism, epilepsy, eye color, to name a few—is defined by more than one gene. A genetic screen could tell you that your baby's eyes will be blue for example, but not the exact shade. The specific outcomes of most multigene interactions are hard to predict with our current level of technology. In most cases, the answer to your question about the likelihood of developing a particular attribute is not a definitive yes or no. Not everyone at risk for prostate cancer develops the disease—some get lucky, some get sick, others get hit by a bus before the cancer develops.

If "nurture" also plays a role in the polygenic trait, the final calculations will be even trickier. For diseases, such external forces are called risk factors. If you have a family history of heart disease, a strict regimen of diet and exercise could help keep your potential disorder at bay, while bacon and video games are only going to exacerbate the problem.

In a rare example of something Hollywood actually got right, the film *Gattaca* deals with this concept fairly well. Just moments after birth, Ethan Hawke's character is given the complete details of his genetic predisposition to a host of common maladies and disorders including obesity, attention deficit disorder, and manic depression. He's granted a 99 percent chance of developing a heart disorder and a mere thirty-year life expectancy, but foresight, willpower, and luck help him beat the odds. Hell, your chances of winning the Lotto are only about one in ten million, but that doesn't mean it can't happen.

IN THE FUTURE, GENE ENHANCEMENTS WILL SWAP IN AND OUT OF YOUR DNA AT THE PRESS OF A BUTTON

In the video game *BioShock*, an undersea colony of scientists, artisans, and engineers begin dabbling in gene modification for personal improvement. Their labors produce gene upgrades that provide the recipients—aptly named "splicers" due to the additional DNA now spliced into their genomes—with some seriously pimped-out abilities like flamethrowing palms or temporary invisibility. What's even better is that these modifications are the very definition of user-friendly—simply walk up to a vending machine and swap upgrades in an instant. Characters are limited only by the total number of upgrades they can have installed at any given time.

No one should have to choose between enhanced strength and a biological TASER. When we humans finally figure out a safe and effective way to insert new code into our DNA, the genes can be inserted nearly anywhere, so long as they don't disrupt something critical or end up in a region that is highly repressed. And since your genome is approximately three-billion-base pairs long while a single gene is only a few thousand, the amount of new DNA is negligible. We'll hardly need to remove obsolete upgrades to make room for new ones. If you happen to tire of a particular upgrade, it'll be far easier to simply shut those genes off—deactivating them instead of removing them. Plan for bioenhancements that can be turned on and off with the flick of a molecular switch.

Installing the new DNA into every cell in your body—or, at least, every cell that needs it—is likely to be the most difficult challenge. Going back in to remove the new DNA from every upgraded cell is going to be a nightmare. That wasp-shooting hand in *BioShock* might sound pretty cool until you send a dangerously allergic uncle to the hospital with a friendly handshake. When you want to turn off that "enhancement" in a hurry, don't try to remove the DNA—render it dormant instead.

TOXIC WASTE AND RADIATION WILL GIVE YOU SUPERPOWERS

While it's true that both toxic waste and radiation are mutagens—agents that can alter your genetic code—"mutation" doesn't necessarily equal "superpower."

Evolution has provided us with some nifty features, such as logical reasoning and appendages capable of fashioning and using complex tools (these may not seem like superpowers to you, but to a bee or a houseplant, they're pretty damn awesome). However, evolution arrived at these traits through a lengthy trial-and-error process. The vast majority of random mutations do practically nothing at all. Of those mutations that actually result in a noticeable change, many are detrimental or even fatal, and only a small fraction are obviously beneficial.

These rare successes create advantageous new genes (like those that might give you a bulletproof duodenum), which are selected for and retained in the population. But mutagens like radiological waste don't make this process more efficient or beneficial—they only make it faster, dumping an unusually large number of mutations on an organism all at once. We don't have hard numbers on this, but we expect that a trip to the nuclear reactor core is far more likely to give you cancer or a third nipple than X-ray vision.

Basically, unless you consider kidney failure and leukemia to be superpowers, you might want to rethink this as a strategy for acquiring abilities beyond the ken of mere mortals.

MUTANT ALLIGATORS LIVE IN OUR SEWERS

We all know the urban legend: Be careful what you flush down the toilet, because it just might mutate into a giant sewer monster. But where did such a bizarre idea come from in the first place? The popular "sewer alligator" myth may actually have some basis in reality. In the 1930s, it was reported that several alligators were sighted in the New York City sewer system, possibly escaped from a ship on its way from Florida. This story was gradually embellished to include giant mutant alligators, spawning several pop culture offshoots including a mutant society living under the streets of thirty-first-century Manhattan in *Futurama* and, of course, the *Teenage Mutant Ninja Turtles*.

But even if your pet goldfish could survive a tortuous journey through your bathroom plumbing, is there any reason to believe that it might subsist in our subterranean waste ducts, growing to epic proportions and terrorizing sanitation workers? Sewers are filled with waste, not with mutagens. There are strict government regulations in place to control the disposal of potentially mutagenic toxins like nuclear by-products and medical refuse, so unless you've also been dumping radon down your drain, there's probably little to worry about.

If you truly desire a nunchaku-twirling mutant reptile, you're better off engineering it yourself. And seriously, stay out of the sewers.

CONCLUSION: MOST OF THE PEOPLE WHO WRITE BOOKS AND MOVIES KNOW JACK ALL ABOUT CLONING AND BIOTECHNOLOGY

Your authors appreciate the arts, especially those arts that include lots of swearing and explosions. Television and film can be soulful and uplifting—they can also be brazenly paranoid, and appallingly, aggressively ignorant. There are a number of subjects about which a person can today be poorly informed without consequence—zeppelin maintenance, for example. Not knowing how your own body functions, on the other hand, leaves you open to more than just ridicule from the better informed. If nothing else, this chapter should have convinced you that the world is full of dangerous misinformation, particularly when it comes to cloning and biotechnology.

Now that we've told you that a number of fun-sounding things are impossible and stupid, it's only fair to toss in a number of awesome-sounding things that are not only possible, they're practically inevitable.

FOUR

BIOENHANCEMENTS

(Self-Improvement That Really Works!)

*I not only think that we will tamper with Mother Nature, I think
Mother wants us to.* —WILLARD GAYLIN

*It has become appallingly obvious that our technology has ex-
ceeded our humanity.* —ALBERT EINSTEIN

he biotech revolution is about much more than just
cloning. Duplicating an existing machine is an impressive technological feat, but reengineering that device is where
the future lies. In the case of humans, evolution has been a
stingy benefactor, leaving plenty of room for improvement.
Sure, it's given us oppressively large craniums, opposable
thumbs, and rockin' genitalia, but think of all of the things it re-

fuses to hand over: wings, flippers, and retractable bottle open-ers. We're not complaining mind you, just thinking ahead.

In the future, there's a good chance that biological enhance-ment will become commonplace. Lasik eye surgery and breast implants are pretty dang popular nowadays, so why not genetic alternatives when they become available? Bioenhancement is simply too revolutionary to be ignored; living without a cellular phone is one thing, but it won't be easy surviving in a world where everyone else is smarter, stronger, and more attractive than you. If the rest of humanity is going to be bioenhanced, then you'd better know what you're up against, especially since enhancing yourself might be the only way to compete. In the future, you'll need to know what's going to be available to you, your clone, and everyone else.

Before we discuss the inevitable technological improvement of the human body via genetic engineering, we feel compelled to note the possible alternatives to biological enhancement.

BIONICS

Lee Majors was awesome, but we'd like to point out that his bionic replacements didn't come cheap; six million dollars back in 1974 would be about twenty-five million today, and that's only the *initial* cost. Unlike machines, our bodies have all sorts of natural repair mechanisms that constantly monitor our health and react when we get sick or injured. When was the last time your car replaced its own transmission, or your toilet un-clogged itself? You might someday be able to buy a titanium skeleton and a telescopic infrared eye, but can you afford the routine maintenance? You can't make do without that artificial heart for very long if it's out of warranty, and most doctors aren't in the habit of giving out loaners while you're between circulatory systems. Are you really ready to commit yourself to a lifetime of high-priced surgeries just so you can see in the dark? Wouldn't you rather invest in a self-repairing biological alternative?

Let's also keep in mind that integrating all of this heavy machinery with your existing biological parts won't be easy. A pair of bionic legs might be great for jumping, but your factory-issued spine is going to have to deal with the landing. Even a simple hip replacement that takes too much of the load will cause the bone it's been hammered into to slack off and become brittle. Have you ever noticed that all of the latest bionic fiction relies heavily on nanotechnology? Well, guess what? Cells have been using nanomachines for millions of years. They're called "proteins." Some upgrade, eh?

Admittedly, if you actually do have twenty-five million dollars lying around, we can think of worse ways to spend it. In fact, we're all for nifty bionic arms that replace the ones lost by amputees, controlled by the same nerves that used to work with the lost limbs. The biosciences can do better.

CYBERNETICS

Perhaps instead of replacing parts of your body with high-tech prostheses, you're more interested in jacking into the Matrix. Promising but dangerous. Reformatting your hard drive after a nasty computer virus is one thing; reformatting your nervous system is something else entirely. Human/computer interfaces are in a very primitive state, and if you want one that does more than move a cursor around on a screen, expect to endure a cybernetic operating system that crashes as frequently as a bumper car.

Besides, cybernetic upgrades can really only lead to one thing: fusion with an AI collective. That's like joining a cult without the benefits of sex or drugs. Do you really want to achieve symbiosis with your World of Warcraft guild?

"OMG u r assimilated. Individuality pwned, n00b."

We didn't think so.

BIOENHANCEMENT

When it comes to improving the human body through technology, the lesson is clear: steel and silicon are no match for cells and tissues. Sadly, while Mother Nature is an extremely successful builder, she generally doesn't take requests. Despite the importance of external factors in human development, we will always be limited by our genetic foundations. You can practice your jump shot every day for the next twenty years, but if you're five-foot-two and physically awkward, you're not going to be the next Michael Jordan. In the past, these inherent genetic traits have been beyond our control, left up to chance and to your parents' questionable decisions as young adults, and we, the offspring, have had absolutely no say in the matter. That's about to change.

The idea of altering a patient's underlying genetic code is commonly referred to as gene therapy, but the practice is still in its infancy. With continued research, ingenuity, and a little luck, however, we'll be able to safely and reliably modify organisms through genetic manipulation by changing the following:

- Which genes are present—Cells are excellent protein producers, but they can only make what their DNA tells them to. For example, botrocetin is a snake-venom protein that causes rapid blood clotting in its victims, but the gene for this protein is found almost exclusively in one particular species of snake. The human genome lacks this specific gene (as well as several others related to venom production), so we're simply out of luck in the "poison spit" department. However, by inserting the code for botrocetin into our own DNA, a human cell could be tricked into producing venom protein. Since viruses have had a lot of practice slipping DNA into human cells, modified viruses are typically used to insert new genes into cells. This must be done with care, because viruses can be haphazard about *where* they insert the new DNA, potentially damaging the genome they're supposed to be enhancing. Other techniques can be used to remove code from a genome, so that we will be able to pick and choose which proteins a cell can or cannot make.

- How much a gene is expressed—The presence of a gene doesn't mean squat if it never gets made into protein; it's just unused genetic filler. In the body, every gene is tightly regulated by a complex combination of proteins and DNA sequences that either promote or repress its transcription to RNA. In nature, this helps cells regulate the amount of protein produced for a given task; but by manipulating these control elements, scientists can reprogram a cell to make more or less of a given substance. Didn't we all know a couple guys in high school who could probably have used a few less testosterone receptors? Extra copies of a gene will also lead to increased production. When a gene is "overexpressed," it's producing more protein than usual. If a gene is "knocked down," it's producing less protein than usual. If a gene is "knocked out," it's producing nothing at all.

· Where or when a gene is expressed—Even the specific spatial and temporal expression of genes is critical in determining the overall outcome of an organism, particularly in early development. For example, a team of scientists recently discovered that they could partially induce the formation of wings in mice by replacing the regulatory elements of a specific mouse gene with their counterparts from a bat. The protein encoded by the gene remained unchanged, but its specific pattern of expression became more batlike than mouselike, leading to elongated mouse forelimbs that resembled precursors to wings. New genes aren't always required to make new biological features.

"Nature versus Nurture" has a nice ring to it, but in the twenty-first century "Nature versus Nurture versus Science" might be more accurate. We have much to look forward to and to be wary of. If you'd like to have better bone structure, a finer musculature, or dorsal fins, genetic engineering may be able to help.

In the case of most genetic enhancements, the technology will most likely become available to your future children (or clone) before it becomes available to you. This may not be fair, but it's easier and cheaper to build something from the ground up than it is to upgrade piece by piece. Consider trying to convert a 1972 Pinto hatchback into a race-ready dragster by individually replacing the engine, chassis, suspension, fuzzy dice, etc. It's much easier just to build a new one from scratch. The same is true for the bodies of you and your clone. Modifying a single cell that will eventually become your clone is far easier than modifying several trillion cells that are already you.

The lesson here is: plan ahead. *Way* ahead. Preferably, before you are born. If you do manage to successfully bioenhance yourself and are subsequently cloned using DNA from one of your genetically bioenhanced cells, be prepared for a clone that shares your upgrade.

Given the diversity of our genetic code and the complexity of our species, the possibilities for tweaking the human body are virtually limitless, but some improvements are more practical than others. Here are a few enhancements that the authors are eagerly awaiting:

INCREASED PHYSICAL CONDITIONING

Regular exercise is not going to guarantee that you'll look good naked, let alone allow you to compete with the bioengineered athletes of tomorrow. We all have that fit-looking friend who eats like a horse and has never seen the inside of a gym. If your postworkout routine includes staring critically at the mirror as you consider the injustice of your genetic predisposition to flab, we're on your side. It's time to level the genetic playing field! A few DNA tweaks here and there and you'll have six-pack abs that beer, junk food, and long hours of lab work under fluorescent lighting can do nothing to soften.

If you're seeking strength and definition, you're looking down the barrel of an enemy protein called myostatin, which signals cells to limit the growth of muscle tissue. Mice that have been genetically engineered to stop producing myostatin have twice the muscle mass of their peers. The same trick works in dogs and cattle. Bully whippets, exceptionally strong dogs, and Belgian Blue, known as "double muscle" cattle, both possess mutations in their genes for myostatin. There's hope for humans as well. In Germany there is a boy who, at age five, had twice the muscle and half the body fat of other children his age, thanks to a pair of inactive myostatin genes. The boy's mother, who had a single functioning myostatin gene, was a professional sprinter. Those of us who are not dogs, mice, or German mutants might benefit from a little genetic reorganization.

If we can create a myostatin-blocking agent, we might be able to develop that eighteen-inch bicep without any tinkering on the genetic level. These myostatin blockers are already being developed as treatments for muscular dystrophy, and once they're on the market you can be sure that athletes will be queuing up. The World Anti-Doping Agency recently placed potential myostatin-blocking agents on its list of prohibited substances.

If double the muscle mass of your friends isn't impressive enough, a mouse that produces no myostatin and has also been engineered to produce unusually high levels of another protein (called follistatin) has *four* times the muscle of an ordinary mouse. Full disclosure: we are talking about some *seriously* freaky-looking mice.

Perhaps you're more interested in a lean, functional build? Mice can be genetically engineered to produce a more active form of a protein called PPARd in certain kinds of muscle tissue. As a result, these so-called "marathon mice" can extend their mousy little jogs over 25 percent longer than their average brethren and avoid weight gain on a diet high in fat. Drugs that affect the PPARd protein have a similar effect. Another gene, adipose, regulates the accumulation of fat, and mice with increased adipose activity tend to be slim.

To combat obesity, we may also have to fiddle with the microbes in our gut, which seem to have undue influence on how we process foods. Just find a health-promoting intestinal population capable of replacing the current residents of your gut, and swallow a sample. When swallowing, try to forget where the sample came from.

Finally, let's be frank. A well-defined musculature can conceal a multitude of geekiness.

PHYSICAL ATTRACTIVENESS

Once you've achieved Olympic-level health, you should consider becoming nauseatingly beautiful as well. Aesthetically pleasing people are happier and more successful than the average person. Believe it or not, the benefits of being tall and pretty have been borne out in study after study. Drop-dead gorgeousness is going to be difficult to engineer at first, but the bioengineering of comeliness *is* going to happen, and soon. If you want to improve a lily, do you gild it or do you breed it?

Currently, our bodies crave sweets and are almost unbelievably stingy with calories, taking every opportunity to put a few away for a rainy day in the First National Bank of Love Handles. When food was scarce and starvation was a leading cause of death, this made sense, but in the developed world today, a beer gut is as vestigial as an appendix. Let's do away with our distressing tendency to pack on the flab and use that energy more productively. It has been suggested that we could engineer our cells to use up excess calories by producing useless proteins, or that we would be better off engineering away our desire for sweets in favor of a healthy high-fiber diet, producing people who disdain doughnuts and cheesecake in favor of bran and celery. Either solution would rob future generations of the energy they'll need to maintain their bioenhanced bodies, and besides, you'll have to pry the ice cream from our cold, dead hands.

Worse, our body's relationship with sweets goes much deeper than taste. Sweet-tasting, low-calorie artificial sweeteners have been linked to weight gain in rats, presumably because the rats' bodies are ramping up to deal with calories that never come, throwing off their metabolisms. The rats end up eating more to make up for the calories they were expecting. On the flip side, mice that lack a gene necessary to taste sweets crave them just the same.

Altering our relationship with food is likely to be a difficult business. Changing what we do with the food once it's eaten, however, could be far simpler. Mice deficient in a single protein—protein kinase C beta—eat more, yet are leaner than mice with normal amounts of the protein. A little genetic tweaking could give us the metabolism of a teenager well into our golden years.

A fit body isn't the only piece of the attractiveness puzzle— ask an NHL defense man to smile if you don't agree. In the near future, your dentist may be asking you to rinse with genetically engineered bacteria instead of mouthwash. Designed to live in harmony with your mouth without producing cavities, this petri-rinse is a quick fix for an age-old problem. Receding gums? Skip the painful surgery and look into stem cell therapies instead.

Height, too, is often a critical factor in attraction between the sexes. We've been using biological factors like human growth hormone to treat cases of short stature for decades. The root cause of such problems is, of course, genetic. *HMGA2*, for example, is a gene with a "tall" and "short" version. Carrying two versions of the "tall" gene correlates with a centimeter boost in height compared with someone who has two copies of the "short" version. We caution you against overdoing it with regard to tallness; too much of a good thing may stress the heart and reduce your life span (the death of Andre the Giant is a sad but classic example of a life cut prematurely short by excessive growth—he was only forty-six when he died).

Still searching for physical perfection? You could also alter your bone structure, eye color, skin tone, and various other features to contribute to your overall look. Think of it as "genetic accessorizing." If you have poor judgment in such matters, please consult an expert. That thrift store T-shirt you thought was so amusing can be easily discarded. Those cheekbones are another matter.

RACE

Bioengineered race is a touchy subject, but one we as a society must grapple with. Biologically, the plus side of mixing genes from various races is that you have potential aesthetic and immune advantages. On the minus side, tissue matching for transplants becomes more difficult, and if you're not keeping a clone handy for transplants (which you shouldn't be!), this may be a problem.

Health-care issues aside, the subject of genetic alteration of race is accompanied by deep spiritual questions. For instance: Will transforming yourself into one-sixteenth Cherokee make you eligible for that scholarship or that tribal casino share? We advise an excellent (and devious) lawyer, and wish you the best of luck.

IMMORTALITY

Want to live forever? Evolution isn't on your side, because Mother Nature only cares if you live long enough to have a few kids. After that, it's out with the old and in with the new. Natural selection is no friend of the individual. Once you pass those prime childbearing years, your body packs it in and slowly self-destructs. Plus, dying is just going to make it that much easier for your future clone to claim dibs on all your stuff.

Given this dire scenario, here are a few things that your local gerontologist might suggest tweaking to increase your life span:

- Reduce the accumulation of mutations. Damage to your DNA accumulates over time and may result in "rude" cellular behavior—cancer, for example.
- Protect yourself from reactive oxygen species. By-products of natural metabolism, these small reactive molecules can wreak havoc on cells and tissues. Preventing this damage with dietary antioxidants is a recurring fad, but we prefer designing a body better able to prevent or repair the damage.
- Carefully regulate the immune system. Autoimmune disease occurs when the immune system mistakes healthy tissue for an invader and attacks it, damaging the body.
- Adopt a youthful pattern of gene expression. As we age, our cells express various genes at different rates. There is some evidence that, by artificially resetting the expression of key genes to mimic that seen in the young, we may be able to increase life span.

Some aspects of aging are within our direct control. For example, eating less (caloric restriction) can extend the life span of animals as diverse as fruit flies and mice. But is it really worth giving up triple-bacon nacho burgers just so we can live for a few extra years? Especially when those years at the end are no fun at all?

If dieting just isn't your thing, then biotechnology is definitely the way to go. In the short term, we may also be able to treat the physical symptoms of aging. Block the activity of the NF-kappa-B protein in aging mice and within two weeks you'll see healthier, improved skin—thanks, presumably, to the skin cells, which begin to act like those in a newborn mouse. Definitely beats Botox.

True life span improvement will probably require a significant genetic overhaul. Scientists have already discovered several genes related to the natural aging processes listed above, either in animal models or by genotyping long-lived humans to hunt for beneficial genes. By manipulating the right genetic sequences, researchers at the University of California, San Francisco, have extended the life of a common roundworm from a mere two weeks up to a whopping five months while maintaining its youthful appearance and vitality. For a human, this would be equivalent to going from a normal life span to one of five hundred years or more. Roundworms have all the luck.

In the end, Methuselah mutations won't make you bulletproof, but they might keep you healthy and physically "young" for a long time (barring accident or disease). In the meantime, you're stuck with moderation and common sense.

HEALING FACTOR

Eternal life sounds a lot less worthwhile if you're destined to end up as a limping collection of scars and old wounds. Injure a human during fetal development and, as long as the damage is not critical, scarless regeneration will often occur. Healing in an adult, however, is comparatively ramshackle, sometimes with function only partially regained. If you're looking to stay in prime shape during your bioengineered immortality, rapid and complete regeneration is a must.

Limb regeneration in starfish and lizards demonstrates that healing in adults needn't always be so "stumpy," and recent studies of a peculiar strain of mice suggest that we mammals have the same potential. These mice, known as the MRL strain, were originally bred to develop lupus so scientists could study the development of that disease. In the course of their work, researchers punched small holes in the ears of the MRL mice to distinguish them from ordinary mice. To everyone's surprise, the holes not only soon healed over, they were repaired without noticeable scarring. The so-called "miracle mice" proved capable of partially regenerating amputated digits and significantly (though not completely) repairing damage to the heart or spine.

Not only is an engineered healing factor useful for recovering from otherwise disfiguring or life-threatening injuries, it's great for practical jokes: "Look Ma, no hands!" (literally). This type of bioenhancement also comes in handy if you're considering a superheroic alter ego (claws optional, adamantium not included).

IMMUNITY TO PAIN

If we can improve our ability to heal injuries, can we also do something about the pain? A mutation in the gene *SCN9A* can render a person completely immune to pain, often with no other evident sensory defects. This may sound like the perfect gene upgrade for a bioenhanced soldier of fortune, but the authors must warn that this level of pain prevention may be counterproductive—pain serves a purpose. Without it, injuries are exacerbated unknowingly by us. Pain is life's way of saying, "Stop doing that, you idiot." As reasonable, self-aware creatures, however, we could afford to trade agony for mild discomfort and still get the message.

IMMUNITY TO DISEASE

An ounce of prevention is worth a pound of cure, and the biotech revolution promises to be rife with new technologies for thwarting disease. We already live in a world where genetic screening makes it possible to select embryos that are free of several genetically determined maladies. Soon, inexpensive personal genotyping will make it possible for you to plan ahead for your most likely Achilles' heel. And with advances in biotechnology helping scientists devise new treatments and medicines, we humans will be able to keep illnesses at bay like never before.

What exactly does the future hold in terms of disease prevention? Let's take a quick run through the list. AIDS? Genetically engineered immune cells could be created that fight

HIV-infected cells more effectively, or resist infection in the first place. Cancer? Mice expressing a modified gene for the tumor-suppressing protein Par-4 become cancer resistant. As an added bonus, they get a several-month bump in life span, which, in mouse years, is not insignificant. Mad cow disease? Well, there's always vegetarianism.

In today's fast-paced society, illness is not only potentially fatal, it's also boring and inconvenient. Who has time for all of these ridiculous ailments? Thankfully, the increasingly disease-proof bodies of tomorrow could provide freedom from a host of illnesses ranging from leprosy to West Nile virus.

In addition to finally liberating us from the shackles of SARS and bird flu, engineered immunity will provide greater opportunities for some of life's little indulgences. Next summer, forget the sunblock and wide-brimmed beach hat. Instead, you can bronze for hours without the hassle of melanoma. And whatever happened to those days of blissful ignorance when smoking was cool and sophisticated? With cancer and emphysema at bay, you can light up those fat, stinky cigars again. Are you tired of Atkins and want to eat deep-fried Twinkies for breakfast? Go ahead, my friend. Your immuno-enhanced body will laugh derisively at heart disease and diabetes. There's also evidence that certain viral infections may contribute to human obesity, and it would be a shame to go to all of that trouble engineering a perfect body and have it go to pot because you caught the wrong cold.

Of course, there are downsides to immunoenhancement as well. If your clone is bioenhanced, this will put a real damper on any biological warfare you might want to wage against your

double. And you can forget about killing him or her slowly with
secondhand smoke. Worst of all, with a bioenhanced immune
system, you'll never be able to call in sick on a Monday again.

IMMUNITY TO ADDICTION

While addiction is considered a disease, it deserves its own
place in our list of bioenhancements due to its unique physio-
logical basis. Besides, blackjack and booze are a lot more fun
than pinkeye and Ebola.

Oscar Wilde once said, "The only way to get rid of temptation is to yield to it." Wilde was a snappy writer, yes, but the
man was no bioengineer. The mechanisms in the brain that
trigger addiction are of particular interest to the tempted. Some
people are more prone to addiction than others; the importance
of family history is well established and the effect of a drug
upon a given person is modulated by his or her genetics. For a
more extreme example, mice with genetically modified dopamine transporters have no preference for cocaine, which seems
to act as a mild suppressant in them, not as a stimulant.

Gene therapy for such addictions—alcoholism for example—
is a definite possibility. Our bodies use a host of enzymes to
break down sweet, sweet ethyl alcohol in the liver. One of the
most important enzymes in alcohol metabolism is aldehyde dehydrogenase, and humans who produce too little of it (or a less
potent form) experience flushing, a rapidly beating heart, and
nausea when they imbibe. How could we use this knowledge to
keep our drinking under control? Scientists now have a breed of
rats with alcoholic tendencies (a breed that ought to go by the

Latin name *Rattus frattus*, but doesn't). When the production of aldehyde dehydrogenase is inhibited in these rats via gene therapy, they go on one of their usual outrageous benders and rapidly come to feel like death warmed over. As one might expect, these rats soon learn to moderate their drinking.

There's also a vaccine for cocaine addiction that is in clinical trials for humans, and a similar vaccine for methamphetamine dependency being tested in animals—both fool the immune system into attacking the drugs, which is a serious buzzkill.

Even the *act* of quitting may be genetically influenced. If you're trying to quit smoking, for example, your genes may not only influence how difficult it will be on you (and everyone around you), they may also influence which method of quitting will work best for you.

If you're prone to addictive behavior, genetic engineering may be an efficient way to get that monkey off your back. Keep in mind, though, that immunity to addiction is also possible for your clones. If you plan to control them with an addictive substance, they'll quickly nip your scheme in the bud and stage a coup. We'll be cheering for them, too, because addicting your clones to crack, meth, or daytime television is a dick move *and* amateurish.

IMMUNITY TO RADIATION

So far, we humans have managed to avoid that whole "nuclear Armageddon" thing, but we're not out of the woods just yet. Low-level radiation is present all around us from sources such as the sun, medical X-rays, and even the potassium in the food we eat. While these doses are typically benign, they may add up over time and play a role in the development of cancer.

Compare yourself for a moment with *Deinococcus radiodurans*, a type of bacteria that can survive heat, cold, vacuum, acid, and five hundred times the radiation required to kill a normal human being. Its capacity for DNA repair is nothing short of incredible, and when repair fails, this bacteria can depend upon extra copies of its genome held in reserve (*D. radiodurans* thinks enough of its genetic material to make backups). Improved DNA repair and storage may prove useful in maintaining our own cells as they suffer the slings and arrows of outrageous entropy.

If this sounds a little far-fetched, there are simpler, less dramatic ways to protect yourself from at least one kind of radiation. People with darker skin are less likely to develop melanomas, so our fairer-skinned readers might want to consider a little pigmentation augmentation when it becomes available. On the other hand, this protection means that darker-skinned humans require more exposure to the sun to synthesize enough vitamin D, which is important for maintaining strong bones and a healthy immune system. So don't get carried away.

SLEEPLESSNESS

Some might remark that an immortal with a healing factor and a boatload of immunities will have plenty of time on his or her hands. We reply, "Not enough." Many hours each day are squandered in slumber, yet various studies suggest that people need even more rest than they're already getting. Who can blame us for coming up short on sleep? It's a time-wasting chore. If it weren't for long-term memory formation, tissue upkeep, and the prevention of psychosis, we wouldn't even bother.

For short-term sleeplessness, a snort from an inhaler filled with the brain hormone Orexin A has been shown to reverse the symptoms of sleep deprivation in monkeys. In the long term, there's a man in Vietnam who claims to have gone without sleep for three decades while suffering no obvious ill effects, and his biology may hold clues that will help us create healthy, continuous wakefulness.

Napping, of course, we'll keep.

It may be some time before we can do away with sleep entirely; so while we're stuck with sleep we might as well make the most of it. Human beings with an alternative version of the gene for adenosine deaminase get more deep sleep than the rest of us. If we have to be in an unconscious hallucinatory state for eight hours a night, it should at least be a *productive* unconscious hallucinatory state.

SENSATION

Even with a bioengineered body that's healthy, fit, and immune to all of life's little irritants, there's still plenty left to tweak. Vision, for example, could be so much better than it currently is (if you're using reading glasses right now, you know what we're talking about). But sharpening our existing senses is only the beginning of sensory bioenhancement. Doing away with hearing aids and contact lenses is all fine and dandy, but seeing an entirely new color wins.

Vision

Pit vipers have a crude version of infrared vision (like the alien in the movie *Predator*) that humans can hope to improve on, because the only thing cooler than seeing in the dark with night vision goggles is seeing in the dark *without* night vision goggles. At the other end of the spectrum (literally), the fat-tailed dunnart (a mouselike marsupial) is capable of seeing extra "colors" in the ultraviolet wavelengths, a feat that our current eyeballs simply can't manage.

Even within the visible spectrum for humans, "color" is a matter of individual perception by our brains. Nearly all humans are trichromats, meaning that we only have three types of cone cells in the retina for color reception—one each for red, blue, and green. All other colors we can perceive arise from combinations of these three. Color blindness is often caused by dichromacy, in which one of the three cone cells is defective or missing. On the flip side, there are tetrachromats who have a fourth type of cone for added color perception "between" the red and the green, which allows them to perceive colors that the rest of us can't even imagine.

While a dichromatic individual will never experience the full range of colors available to trichromats and will often have problems with color-related tasks such as interpreting traffic signals and reading certain Web pages, a true tetrachromat will have a richer color experience than the rest of us, because she is capable of distinguishing between colors that look identical to a typical trichromatic human.

The bottom line is, we're surrounded by sights that our biology blinds us to. We will be upgrading to dodecachromacy as soon as our HMO acknowledges our spectral impairments.

Olfaction

Our sense of smell may seem like a relatively useless thing to bioenhance, but that's only because we don't know what we're missing. Estimates vary wildly, but a dog's sense of smell may be up to a hundred million times more sensitive than a human's, allowing specially trained canines to detect narcotics hidden in airport luggage, or even to smell cancer on a human patient's breath. Other species such as the silkworm moth have such an acute olfactory sense that they can smell a *single molecule* of moth pheromone in the air. The authors, on the other hand, are lucky if they can smell a house on fire. A bioengineered sense of smell could come in handy for anything from crime fighting to cow pie avoidance.

Enhancing your own sense of smell is fine, but what about the scents you're giving off to others? Maybe it's time to alter your own personal body odor (i.e., your pheromones) to make it more appealing to potential mates. Women are drawn to the smell of a man with a major histocompatability complex (MHC) genetic profile dissimilar to their own—MHC is a gene family that is critical to the immune system, and if your mate's MHC genes are similar to yours, there's a chance you're engaging in a little recreational inbreeding.

Of course, there are simpler methods for attracting humans via smell: strawberry- and cocoa-scented genes are especially popular with the ladies. Sure, we get some odd looks from the guys, but it's *so* worth it.

Taste

The five basic tastes are bitter, sour, sweet, salty, and umami (a Japanese word for "savory"), but not everyone experiences them the same way. Supertasters are capable of experiencing certain flavors differently, or with an unusual intensity as a result of genetic variation. Combine a few supertaster genes with increased olfactory ability and a simple meal could become a symphony of flavors or an overly flavorful and inedible mess. You don't want to become too picky, especially if your other enhancements require you to consume ten thousand calories a day.

Equilibrioception

It would be a waste to sculpt a perfect body stacked with musculature that lumbers about making you look like a buffoon. A powerful body without a good sense of balance (equilibrioception) is like a car with a V12 engine but no power steering. If you want the strength of a linebacker with the moves of Jackie Chan, you're going to need the whole package.

We maintain our equilibrium with the help of a structure in the inner ear, which contains fluid that tells the body which direction is "down." There's potential to improve our native balance-keeping system, but that's not our only option. Some marine invertebrates use tiny stones instead of fluid, and many animals have tails that help them maintain their balance. Sure, your bioenhanced clone might look a little silly with a lemur tail sticking out his backside, but you won't be laughing if you have to fight him on a rooftop or a tightwire. Hey, it could happen.

Proprioception

Proprioception is related to equilibrioception, but subtler. To perform gracefully, the body needs a sense of where its various bits are located. This "body sense" is critical in developing hand-eye coordination and muscle memory. When the police ask you to close your eyes and touch your nose, they're testing your proprioception. Don't neglect this sense when considering your suite of bioenhancements! Poor proprioception will prevent you from taking full advantage of many of the other physical enhancements, while improved proprioception will have you traversing security laser grids while blindfolded.

Echolocation

Bats, dolphins, whales, and lawyer-vigilante superheroes all use echolocation, or biological sonar, to sense their surroundings, so why not you? We admit that the squeaking can be undignified, but not as undignified as bumping into something in the dark. Studies have shown that human neurons are unusually adept at discriminating between small changes in sound frequency compared with those of other mammals; several blind people have already developed a primitive form of echolocation, suggesting that true echolocation in humans is at least possible. A little genetic tweak to increase the range of our hearing might help. A larynx tuck might not hurt, either. In certain species of bats, for example, a gene contributing to vocalization has been extensively tweaked by evolution to help the bats make sonar-friendly chirps. Take care choosing this enhancement. If it gets too popular, you'll be listening to everyone else's high-frequency bleating all night.

Magnetoreception

This unusual sense allows an organism to "feel" magnetic fields, a critical attribute for migratory animals who are unable to carry a compass and can't afford a GPS. Birds do it. Bees do it. Even some bacteria do it, with the help of magnetite crystals in specialized organelles called magnetosomes.

You might find such a "biocompass" useful while hiking or for getting the most out of expensive audiovisual equipment. Careful: too much sensitivity and you'll never be able to endure an MRI scan again without screaming.

CHIMERISM

The chimeras of myth are imaginary creatures made up of parts of multiple animals: fantastic mix-ups of various limbs, torsos, and heads. Real-life chimeras are animals whose cells do not all originate from a single zygote; and unlike griffins, unicorns, and mermaids, they actually exist. To create a chimera, two or more genetically distinct sets of cells must combine to form a complete organism, resulting in an animal that consists of a patchwork of cells and organs from different lineages. This may sound like something straight out of science fiction, but every recipient of a bone marrow transplant could reasonably be called a chimera. A bone marrow recipient receives stem cells from the donor that take root in his or her body, replacing cells damaged by disease or chemotherapy. Careful matching is required because the cells from the donor remain genetically distinct from the rest of the recipient's body and are subject to rejection by the immune system.

Not all human chimeras are created deliberately. Imagine being told that you were not the genetic parent of your child. For fathers, this is not inconceivable, but what if you actually *gave birth* to that child? A woman named Lydia Fairchild lived out this exact scenario: according to a DNA test, she was not the mother of two of the children she had carried in her womb. Pregnant with a third child, she had blood drawn from the baby at birth—the result was the same, and was confirmed by multiple labs. It was then suggested that perhaps Lydia was a chimera—in the womb, embryo Lydia had fused with another fertilized egg, and the genetic material of both eggs persisted in the adult Lydia. Tissue taken from her thyroid matched her children, proving that she was their mother, but blood, hair, and saliva samples indicated a mismatch. Hers is not a unique case—a teacher named Karen Keegan, looking for a kidney transplant match, was told that of her three adult sons, two were not her own. She, too, was eventually discovered to be a chimera.

It's possible that most mothers are chimeras of a sort. Stem cells from a developing fetus can take root in a mom and live in relative harmony with her. In humans, these cellular hangers-on have been known to persist for decades. In mice, fetal stem cells have even been observed slipping into the mother's brain.

By leaving a little of yourself behind, you quite possibly turned your own mother into a microchimera. We're not judging. It goes both ways: you may also have a little bit of your mother in you. Mother-child chimerism is probably unavoidable and could be the cause of some autoimmune disorders; yet another reason to give that immune system of yours a tune-up.

The careful combining of human embryonic stem cells with fetal mice can produce chimeras containing cells from two different species. Mice with human cells in their craniums have been produced by this method. These mice aren't superintelligent, but they do make superior study models for neurodegenerative diseases. Mice with human immune cells or human liver cells are used in HIV and pharmaceutical research. Sheep and pigs with human organs produced by similar means may become handy replacements for transplant donors (once the bugs have been worked out). In his 2006 State of the Union address, President George W. Bush called for legislation against the creation of human/animal hybrids, but we like to think of chimeras as animals that nature wanted to make but never got around to.

There are many advantages to setting up a living arrangement with tissue that is not your own. Perhaps your genes have a genius for gray matter and muscle mass but have whipped up a terrible liver. Why not let your DNA do what it's good at and hire out the construction of those other organs to borrowed DNA that's better equipped for the job?

The advantages don't end there. Anyone who has seen *CSI* knows that it's just about impossible to avoid leaving behind evidence of yourself everywhere. So, how can you prevent leaving a trail of your genetic material wherever you go?

First, you'll need to keep your home or apartment as clean as a surgical suite. Glasses and utensils naturally drip with saliva after use and therefore must be sterilized (this makes eating out a no-no). Avoid leaving blood or tissue samples with your doctor; if necessary, tell medical staff that every piece of you must be buried with you for religious purposes. Keep your hair and nails short and cut them yourself, burning the clippings afterward. Better yet, consider purchasing an industrial incinerator for thorough elimination of all your disposable items. Wash your clothing daily with bleach—and no sending out the dry cleaning! Remember, if you have a clone of your own (or a twin), he or she is going to have to follow all of the above rules, too.

If these lifestyle changes sound like a downer, remember that thanks to chimerism, there's no reason why your skin and hair need to share your DNA. A chimera can keep all of its favorite nucleic acids in its less-exposed organs and farm out its skin and hair to a ringer. It won't be easy—skin is the body's largest organ, and replacing yours with what is effectively someone else's would leave you open to some very painful operations, plus the possibility of immune rejection.

If you sort out these technical issues and decide to mask

your genetic profile using chimerism, we suggest using celebrity hair or skin for maximum impact. Besides, famous people tend to have nice hair and clear complexions.

ANIMAL TRAITS

Humans won the genetic lotto and rule the planet with language and intelligence, but that doesn't mean we got *all* the good stuff. If you want your next child to have a pelican beak or a rhinoceros horn, you could try mixing the human and animal cells together. The problem with chimerism, however, is that you never know exactly what you're going to get. In pop culture, chimeras are usually depicted as having distinct traits from two or more organisms. In reality, getting exactly what you want is hard to do. Some chimeras are created by fusing two embryos early in development. The new chimeric embryo contains stem cells from both organisms which develop in parallel to give rise to a single animal. This is a quick and easy way to generate chimeras, but it's the equivalent of a cellular crapshoot. The stem cells from each original embryo have a roughly equal probability of developing into the various body parts of the new chimera. In the lab, this is (usually) not a problem if the chimera is a mixture of two embryos from the same species. A human/human chimera will simply develop into a human, albeit a rather unique one with tissues and organs that contain two different sets of DNA.

When you start combining different species things get messy. It's much easier if you want a "mostly mouse" with relatively few human neural cells, which can be introduced by injecting human neural stem cells into a newborn mouse. By waiting for the mouse to develop to a certain degree, you've given the mouse cells a clear head start. By using neural precursor cells, you've made it more likely that the injected human cells will form neural tissue in the mouse brain instead of in other organs that you'd rather keep completely mousey.

Let's say you want to make something distinctly and obviously multispecies, like a mermaid. You start with a human embryo and a fish embryo and—using a little freshman biology—you fuse them together. You take this chimeric embryo and implant it in a highly adventurous surrogate mother. Nine months later, your mermaid baby is born. Congratulations! It's a girl . . . fish? If you're unbelievably lucky, maybe your chimera has the head and body of a woman and the tail of a fish. When you first merge the two embryos, the fusion creates a critical starting point for the development of your chimera. The resulting tissues and organs that make up your "mermaid" will be determined mostly by the locations and interactions of the original fish and human cells within this chimeric embryo. Because the initial fusion is largely random, it's doubtful that your chimera will be a perfectly bisected fish/girl. Instead, the most likely outcome is some sort of girl/fish mosaic—an organism with a random mixture of fish and human body parts. Wait, we take that back. The probability that your human/fish embryo will develop into any kind of viable organism is pretty low.

While "directed chimerism" still has a few hurdles to overcome, it may still be possible to enhance humans in a less haphazard way. If the traits you're interested in can be reduced to the addition of novel genes, genetic modification will be more likely to result in an upgrade than a "cross your fingers" chimera. Such modifications are likely to be somewhere between preposterously complex and impossible, but, oh, so much fun. Here are a few to get you started:

- Porcupine spines: rough on predators (and couches).
- Chameleon skin: nude, yet camouflaged. For the shy exhibitionist.
- Gecko-grip hands: good for scaling walls or opening stubborn jars.
- Worm, lizard, or starfish-limb regeneration: because a gecko-grip hand is only useful when it's still attached to your body.
- Rattlesnake venom: finally, your dentist is more afraid of you than vice versa.
- Rhino hide or crustacean exoskeleton: for those with bioenhanced physical strength only!
- Plant photosynthesis: reduce your carbon footprint by absorbing carbon dioxide as you tan.
- Electric-eel bioTaser: potentially useful for self-defense and CPR, though it's likely to be hell on your cellular phone.

Not all animal bioenhancements are as exciting as the ones we've listed above—no one wants to look at your new, brightly colored primate posterior with skunk scent glands. Special caution should be used when contemplating the use of genes recovered from extinct species. Tempting as it might be to splice yourself with Tyrannosaurus rex genes, after millions of years they're not likely to be in the best of shape. If by some miracle they *do* function, consider how hard it's going to be to drive with those tiny arms.

INTELLIGENCE

Our readers are already of above-average intelligence, but we caution you: don't get cocky. If you let brainpower slide in your quest for superpowers, you could be left in the dust of the mentally enhanced masses. Besides, an improved intelligence will help you develop and refine your other enhancements.

Before we can improve intelligence, however, we must define it. Intelligence is a nebulous concept, encompassing a variety of different talents that may each be influenced by its own set of genes. A gene that predisposes you to advanced mathematical thinking, for example, may have little or no effect on your ability to learn a second language—and the effect of a single gene is likely to be negligible anyway. Real improvements in intelligence are probably going to require a coordinated effort by many genes, or some fundamental reengineering. It's possible that we'll stumble upon a trick or two, however—in mice, at least, the overexpression of a single gene called *NR2B* leads to more rapid learning with better recall. These genetically engineered "Doogie Mice" (named after television's Dr. Douglas "Doogie" Howser) do not yet have their own coming-of-age sitcom, but they've given us at least one potential link between genetics and intelligence.

In humans, there's a correlation between a history of poor decision making (or, possibly, risk-taking behavior) and a mutation that leads to decreased numbers of D2 receptors in the brain. These receptors are involved in how our brains react to changes in dopamine concentration, and dopamine—among many other things—helps us learn. Associating a pleasant dopamine buzz with a decision makes it more likely for that decision to be repeated in the future. Thanks to brain chemistry, you are your own Pavlov, but genetic engineers may soon be the ones ringing the dopamine bell.

Even if bioenhanced intellect becomes possible, there are numerous aspects to consider. Simply *measuring* intelligence is a difficult task fraught with contention. Assigning a value to intellect is no mean feat, as any social scientist (or high school teacher) will tell you. Psychologist Howard Gardner breaks intelligence down into various components, each of which we might someday be able to manipulate, including:

- Logical: Capable of reasoning consistently without falling prey to fallacy or paradox. Think of Vulcans or the ancient Greeks. On second thought, just think of Vulcans. Some of those ancient Greeks were *crazy*.
- Linguistic: Adroit with languages. This would presumably include poetry, though not anything either of your authors have ever read. There are also genetic profiles that seem to make it slightly easier to learn tonal versus nontonal languages and vice versa (if that Chinese language tape is giving you trouble).
- Spatial: Able to visualize shapes and manipulate them

mentally. Useful for considering architecture, M. C. Escher's work, or for playing Twister.

- Musical: A sense of rhythm, tone, and pitch. If you do karaoke regularly, chances are you should look into improvements in this area.
- Kinesthetic: A strong sense of one's body, especially in motion. Finally, you will be able to understand why you can't dance.
- Intrapersonal: Self-awareness. This is different from being aware of your clone, remember.
- Interpersonal: Suavity, charm, and savoir faire. Getting along with your spouse, co-workers, or clone will save you no end of trouble.

While self-control is not usually considered a component of intelligence, it's hard to argue against it as an ingredient for success. Recent evidence suggests a surprisingly strong connection between genetics and what psychologists and neuroscientists call "executive function." Think of the part of your brain devoted to executive function as the ant that chides the more-grasshopper areas of the brain, quieting their cries for immediate gratification in favor of long-term goals.

Look for synergies between your other enhancements and an upgraded intellect. If you opted for better musculature, an improvement of your kinesthetic intelligence will let you take full advantage of your new body. Heightened interpersonal skills will allow you to charm those drawn to your biologically sculpted physique instead of scaring them away with your God-given awkwardness.

Or, if you're willing to take a risk, clone yourself many times, enhancing one aspect of intelligence in each clone—instant sitcom and/or existential crisis.

MEMORY

An intellectual boost is only as good as your ability to retain and recall all those new insights. Whether you'd like to lay claim to an eidetic memory or simply become better at remembering the names of people you meet at parties, memory enhancements are a must. Just as a computer with a speedy processor and a tiny hard drive is only marginally useful, so too is an intelligent brain with a limited capacity to commit things to memory.

Memory is a difficult attribute to pin down. Like intelligence, what we think of as memory is a complicated process involving different parts of the brain. One common way to distinguish the types or stages of memory is to consider short-term memory and long-term memory discretely, but even this is a drastic simplification of what really happens inside our skulls on a regular basis.

Despite the intricacy of the system that allows human beings to remember where we left our car keys (sometimes), there are good reasons to believe that real improvements can be made. In humans, there are two common variations of a protein called brain-derived neurotrophic factor, or BDNF. These variants are called "Val" and "Met." Not only do people with the Val version of BDNF score higher on standardized memory tests than people with the Met variant, they also demonstrate more brain activity during the encoding and retrieval of information.

This makes Val a potential treatment for forgetting where you left your car in a mall parking lot.

The second most serious problem regarding memory is how it begins to fail with age. (The authors have forgotten what the first most serious problem is.) Luckily, stem cells offer a possible alternative to senior moments. Mice with induced brain damage in the cells of the hippocampus—a region in the brain that helps control recall—suffer from memory impairment, but subsequent injections of neural stem cells completely reverse the loss. It's also possible that your biggest problem with memories could be having too many of them. We caution you that a *perfect* memory is not necessarily your goal. At least one remarkable mnemonist, the Russian journalist Solomon Veniaminovich Shereshevskii, would write words down and then burn the paper in a desperate attempt to forget them. He claimed that remembering too much was confusing.

Finally, you may have seen something you would very much prefer to unsee. By manipulating levels of a protein called αCaMKII in the brain, researchers have been able to selectively "edit" memory recall in mice. It's not a *Men in Black* "neuralizer" by any means, but if you're desperate to forget last Tuesday's blind-date fiasco there may one day be a biological option.

EMOTION

Just as mood swings can be brought on by low blood sugar, our emotions can be influenced by the hormones in our blood and also by our genes. A family history of clinical depression may indicate a genetic predisposition to the disease, and variations in genes that control the body's reaction to stress hormones can reduce one's chance of developing posttraumatic stress or depression. Many pharmaceutical treatments are thought to work by adjusting the levels of various neurotransmitters, the chemicals that transmit signals in the brain, though more recent research suggests drugs like Prozac may also protect neurons and promote their growth. Whether it's the neurotransmitters or the neurons themselves in need of adjustment, there's no reason why a bioengineering approach won't someday be able to do the job.

Treatments for depression will probably be the first bioenhancement of our emotional states (based on the clinical need alone), but they'll hardly be the last. With millions of people undergoing psychological treatments every year, it's obvious that our emotions can sometimes get the best of us; and while love is a many splendored thing, obsession is a one-way ticket to a restraining order. Emotional modification may someday be your ticket to a more balanced psyche.

If your fears are dragging you down, for example, gene therapy may offer hope. Identical twins are more likely to share phobias than fraternal twins, indicating that some of the tendency toward fearfulness resides in their genes. When designing genetic adjustments to avoid paralyzing anxiety, keep in mind that an overabundance of heedless bravery would not necessarily be a good thing. One man's immunity to fear is another's addiction to risky behavior (street luge! Woo!) and no amount of bioenhancement is going to protect you from your own spectacular lack of judgment. For instance, scientists can create mice that are as unafraid of cats as humans are, but *our* not fearing cats is rational—for mice, cozying up to a feline can be fatal.

On the other end of the spectrum, our more blissful emotions can also be manipulated through science. Humans who have recently fallen in love and are still in the "honeymoon phase" have an increase in the amount of nerve growth factor (NGF) in their bloodstreams; the more passionate the feeling, the higher the NGF level. After a year or so in a relationship, blood NGF sinks back down to the amounts you'd see in single people or settled couples.

We don't recommend dosing yourself with NGF to maintain the rush of new love well past the paper anniversary. Nerve growth factor—as its name implies—is not simply "the love chemical." It's involved in a number of biological processes, so it might not be wise to attempt turning yourself into a hopeless romantic by overexpressing it. On a personal note, we authors don't want to deal with any more cloyingly saccharine couples than we already have to.

Even our choices in love can be modified with a little chemistry. Take prairie voles, for instance: most are monogamous, but a few males known as "wanderers" just can't be tied down, baby. Pump these prairie lotharios full of a hormone called vasopressin and they go on to form monogamous pairs. Block the vasopressin and prevent it from functioning, and goodbye monogamy. The hormone is part of the reward system of the brain, supposedly the part of the system that causes the warm fuzzies in voles. A vole that strongly associates the pitter-patter with his mate tends to stick around.

So, what makes the wanderers different from the typically monogamous prairie voles? It turns out that they have a different distribution of a certain receptor that recognizes vasopressin in the brain. This distribution may be throwing off the love signal, making it more difficult for these wanderer voles to settle down with a single mate. A related species, the meadow vole, is normally promiscuous. Boost a meadow vole's expression of the vasopressin receptors in a particular part of the brain, and it settles down with a single mate and trades in the sports car for a minivan.

Most of our readers are not prairie voles, so we don't necessarily recommend postcoital vasopressin injections for your fickle significant others. It's immoral, illegal, and no amount of warm fuzzies is going to overwhelm potential mates to such a degree that they don't wonder what was in the syringe you just jabbed them with. Furthermore, some hypothesize that—in prairie voles at least—vasopressin-induced monogamy has more to do with memory formation than twitterpation. The monogamous male voles may be better able to remember where another male's territory is, and less likely to blunder into a fight with another vole's boyfriend. In humans, there are indications that males with a certain version of the gene for the vasopressin receptor tend to be less committed in their relationships, while other genetic variations in the same gene have been linked to traits as dissimilar as generosity and (believe it or not) creative dance performance. Proceed with caution: Our emotions are highly networked with other brain functions, and it's difficult to alter one emotional state without inadvertently raising a neurological ruckus. When it comes to human emotion, no gene is going to guarantee the result you're looking for, and any attempt is likely to have unintended emotional consequences.

Toying with other peoples' emotions without their consent is offensive behavior, whether it's done socially, chemically, or genetically. If that doesn't convince you to approach emotional bioenhancements with a little trepidation, keep in mind that they're also damnably difficult and laden with potential side effects.

MORALITY

We are social animals, and it's no surprise that much of our brains are devoted to dealing with other human beings. Getting along with those around us was (and is) a good way to survive; being ostracized was a good way to make sure that your genes were consigned to nature's rubbish bin. Guilt, for example, can prevent us from acting in a way that will harm or offend others. Righteous indignation, on the other hand, can convince those around us that if we catch the bastard stealing lunches from the break room there will be *retribution*.

While certain aspects of morality vary between cultures, others don't. Homicide isn't going to get you invited to many parties no matter where you grew up. The moral codes of a pacifist hippie vegan and a bible-thumping preacher may seem as different as night and day, but they do share certain rules, like not having sex with their siblings or farting on the *Mona Lisa*. If our biology includes universal moral instincts, will those instincts change if the underlying biology is changed?

Twins studies have shown that certain types of antisocial

behavior in children—early warning signs of psychopathy—have a genetic basis. If there are combinations of genes that predispose human beings to lack empathy and remorse, there ought to be other gene combinations that predispose us to play well with others. By uncovering the biological groundwork of morality, scientists may be able to bend it to their wills. Of course, raising a human being to have principles in a nurturing environment will usually do much the same thing at a fraction of the cost.

Morality bioenhancement carries with it a unique potential for questionable behavior. Just as genetically enhanced virtue could make you a natural arbitrator or guardian, a sociopath-by-design would possess an attenuated empathy preparing him or her for a successful career as an assassin or a lawyer. Bioengineered morality can also be extremely awkward: you may design clones to be morally superior to yourself only to have them tell you that you shouldn't have done that.

CONCLUSION: THE HUMAN BODY IS A UNIQUE FIXER-UPPER OPPORTUNITY

As machines, humans are fairly impressive, but if we had to grade the design we would definitely give it a "room for improvement." While some of the enhancements in this chapter may still be decades away, we hope to see many of them early enough in our lifetimes to enjoy. Whether you want to be smarter, prettier, or genetically battle-ready, your body will be yours to enhance—and so will your clone's. If you're forced to compete with the beautiful, athletic, and ultraintelligent clones of tomorrow, you'll need to choose your own personal improvements wisely.

We're as eager as the next person to experience the full range of human biological enhancement. Yet, as scientists, we're painfully aware of the inexorable law of technology: our kids are going to have much cooler stuff to play with than we ever did. If you're understandably impatient for the benefits of biological augmentation, we suggest looking for a less ambitious project than a total reinvention of the self—try a total reinvention of the tomato instead. Lower life-forms were put on Earth for a purpose: they make excellent test subjects.

FIVE

A STARTER KIT FOR PLAYING GOD

(Fiddling with Lower Life-Forms)

There's nothing like millions of years of really frustrating trial and error to give a species moral fiber and, in some cases, backbone.
—TERRY PRATCHETT

I have created things that will change the world for the better. For example, here is a monkey with four asses.
—DR. ALPHONSE MEPHISTO, *SOUTH PARK*

hy do most scientists begin an experiment using cells or animals as subjects instead of humans? Simple: the first thousand or so times, the experiment flops, so be patient. Leaping headlong into personal bioenhancement or cloning without a firm grounding in the basics is bound to slap you with cancer, a metabolic disorder, a cranky genetic duplicate with an unusual number of toes, or (at best) an expensive failure. Practical experience with genetic tinkering is an excellent way to familiarize yourself with the promises and pitfalls of the biotech revolution, but by all means begin with lower life-forms—bacteria, plants, or animals—that can't sue.

In this chapter, we introduce some of Earth's greatest biological test subjects and the promise they hold for our bioengineered future. Pay close attention, and you might even learn some tricks of the trade for designing your own food, pets, and more. Besides, medical research is notoriously expensive. A lucrative lower life-form could provide the bankroll you're going to need if you want to branch out into *Homo sapiens*. Who knows? Maybe your adorably bloodthirsty hammerhead koala will be a useful ally when your clone tries to infiltrate your Jacuzzi.

VIRUSES

For some, viral research means only one thing: biological warfare. Yes, viruses may someday be reengineered as ultra-infectious, megalethal bioweapons. If this type of work sounds like fun to you, please set this book aside and shun the sciences in favor of a position in which your antisocial tendencies can be put to safer use. We suggest politics or the slaughterhouse industry; but, really, any field where dropping a test tube won't turn Seattle into a mass grave will do. Take a hint from movies like *Resident Evil* and *28 Days Later*—the first victims of any bioengineered plague are usually the scientists who designed it.

Not all work with human pathogens involves building

newer and deadlier viruses, however. Consider the 1918 in-
fluenza epidemic that killed more people than World War I. To
discover why this particular strain of flu was so spectacularly
deadly, scientists took tissue samples from victims who had
been frozen, then attempted to get viable flu virus from the
samples. They failed—the virus was present, but its genetic material was in tatters. Their only option was to sequence the bits
that remained, reconstruct what the virus would have looked
like, and then rebuild it. The resurrection of potentially fatal
viruses is a very poor starter project for you, but genetically reconstructing a deadly yet unavailable virus gives us a chance to
figure out why it was so dangerous in the first place. Anticipat-
ing and preparing for future outbreaks requires an understanding of past pandemics. When a disease is no longer around,
getting a handle on it requires some fancy biotechnology, and
this sort of work is excellent practice for reconstructing the
woolly mammoth genes you'll need to protect chilly elephants
or grow yourself a really kick-ass coat for the winter.

If you're just starting out, it's safer to work with a virus that
has no interest in making you ill. Bacteriophages are viruses
that shun humans in favor of bacteria, and if you come down
with a bacterial infection, then the enemy of your enemy can be
your friend. Phage therapy seeks to combat infection with more
infection: a bacterial infection is itself infected with a bacteria-
killing virus. Create a viral alternative to antibiotics and it'll be
used in hospitals and hand soap in no time flat.

Remember, viruses may be tiny, but they're not to be under-
estimated. Gene therapy and medicine may come to rely
heavily on these natural "genetic engineers."

Suggested projects:

- Virophages—Viruses that can fight other viruses. This phenomenon may be extremely rare in nature, but that just makes it more fun.

- "Joy" virus—In the movie *28 Days Later,* most of the London populace turns zombielike with a sickness known as "Rage." If you're going to invent new infections, how about one that makes us all cheerful and polite instead of viciously psychotic? Even a "mild disinterest" virus would be better than a "Hell hath no fury" one.

- Special delivery—Viruses are very good at smuggling DNA into cells, but often at the expense of that cell. A designer virus that can safely and efficiently drop its genetic payload into human cells would be a handy tool for gene therapy.

MICROORGANISMS

You could do worse than using a microbe for an experimental subject. Quick and easy to grow, hard to kill, and simple enough to play with, scientists have been meddling with *E. coli* and its brethren for decades. One good incentive for studying bacteria is that many species can be easily coaxed into internalizing bits of DNA from their surroundings, usually by starving the bacteria or otherwise stressing them. For bacteria, this is probably the evolutionary equivalent of pushing random buttons when you know things can't get much worse—maybe, just maybe, there's a piece of DNA out there that will help the microbe survive. Scientists can take advantage of this phenomenon and use it to "force-feed" specific DNA sequences (genes, etc.) into a bacterial cell. If you're dealing with a species that never gets that desperate, specialized solutions or powerful electric fields can be used to coax DNA into the bacteria.

Let's say you want microbes that glow under a UV light like a jellyfish does. Simply find out which protein in the jellyfish is responsible for the glowing (it's the aptly if uncreatively named "green fluorescent protein"), then identify the gene that codes for that protein. Package the corresponding DNA into a plasmid, which is basically the gene for the glowing protein plus some extra bits, including DNA code that fools the bacteria into expressing the gene. Stuff that plasmid into a bacterium and you have yourself a particularly decorative microorganism under black light. Yet, designer bugs can do more than just glow; most of the medical insulin sold today is made in one of the world's smallest (but most numerous) factories: *E. coli*. While not all proteins can be usefully produced in microorganisms (owing to protein size or structural complexity), they possess a remarkable degree of flexibility.

Sometimes the new genes we put into bacteria aren't nearly as interesting as the ones that are already there. Microorganisms have been around for a few billion years, which is plenty of time to evolve methods for doing some pretty extraordinary things. "Extremophiles" thrive in acid or severe temperatures. Some bacteria can even survive in a vacuum or endure dangerous levels of radiation for short times, while others can munch happily on uranium or modern antibiotics. Nature has already provided us with a staggering number of genes—it's up to us to find them, understand them, and then play mix-and-match.

Combine a bacteria that can resist radiation with a gene (from another bacteria) that confers resistance to dangerous levels of mercury, and you might have a bug that can help clean up radioactive mercury-contaminated waste sites. Just the thing to get rid of that box of thermometers you bought on eBay from ChernobylSeller137 (seriously, what were you *thinking*?).

If evolution hasn't yet produced an adequate solution to your particular problem, give it a little push via directed evolution. Let's say a particular species of bacteria performs a certain task moderately well—breaking biomass into biofuel, for example. Take one of the genes responsible for turning that biomass into fuel and induce various mutations in it, producing many different versions, each of which can be slipped into a microbe to see how well it works. Give each microbe an opportunity to chomp on some biomass, and keep an eye on which versions of the bacteria are new and improved. Bacteria that aren't up to snuff get snuffed, while those that thrive contain potentially valuable mutants.

Can't find a bacteria in nature that performs as you'd like it to? You may be able to finagle an existing species to suit your needs. Scientists are putting together increasingly complicated combinations of DNA, or "parts," that can be used to modify bacteria. Simply find or build a part that fulfills your need, get ahold of the DNA that constitutes that part, and—properly installed in a bacteria—the part becomes a fully functioning system that can tell your bacteria when to glow, when to die, or what to synthesize. With the decreasing costs of molecular biology and the increasing amount of information online, we're not far from the era of open-source garage-kit biology. There's also the potential to start from scratch. Adding new genes to bacteria can be a rich and rewarding experience, but inventing an entirely new bacterial species means you get to name it. (Anything vaguely Latin-sounding will do.)

Suggested projects:

- Homebrewer's yeasts—While unusual flavor-producing yeasts are a possibility, we'd be happy with one that doesn't produce unpotable swill in the hands of an amateur brewer.

- Engineered probiotics—Replace the flora found in your gut with bioengineered bacteria that will fix a host of problems (such as lactose intolerance) and make your silent ones less deadly.

- Extremeophile challenge—Design a microbe capable of surviving ludicrously harsh conditions. Pit your microbe versus others in microscopic combat. Two bacteria enter; one bacteria leaves.

FOOD

Viruses and bacteria are extremely useful in the lab, but they can be a drag to work with on such a tiny scale. Let's move to something a little more hands-on (and delectable).

In the wild, tomato plants can produce fruits 1/1,000th the size of a domesticated tomato. The difference likely has something to do with a mutation in a gene called *fas*, which controls the number of compartments per fruit. Wild tomatoes, lacking this mutation, are relatively puny compared with what you'd pick up at the farmers' market. This mutation (and others) are thought to have arisen naturally during the domestication process. Grow enough plants and eventually one will pop up with fruits that are larger than the others. The farmer, knowing a good thing when she sees it, cultivates the unusually productive plant. Repeat over many seasons and the farmer will have produced, via selection, a plant genetically distinct from the one she started with.

Obtaining a plant with the characteristics you want needn't be a time-consuming process. There's no need to wait for a desired trait to appear naturally as your crops get jiggy with it, plant style. We already have an arsenal of laboratory tools to introduce specific genetic changes.

The term "Frankenfood" is commonly applied to genetically modified (GM) fruits and vegetables by opponents of genetic manipulation, but the metaphor suffers upon reflection. The patched-together Frankenstein's monster has more in common with the traditional horticultural practice of grafting one plant to another than it has with genetic alteration, which also requires no lightning and rarely involves shouting, "It's alive!"

The first genetically modified foods were crops altered to increase resistance to insects or pesticides and extend shelf life. Specialized rapeseed plants now survive the use of insecticides that would kill their unaltered brethren, while a unique variant of sweet corn has been modified to generate its own insecticides. These alterations make the crops easier to cultivate, but otherwise the product is unchanged. Plants can also be optimized to produce biomass for alternative fuel sources, or to be more nitrogen-efficient, reducing the need for nitrogen-based fertilizers.

There are likewise opportunities to make food healthier. Inserting genes from snapdragon flowers into tomatoes produces a fruit that is high in antioxidants (and purple). Rice naturally produces beta-carotene, a nutrient with antioxidant properties, but not in the edible part of the plant. "Golden rice" has been engineered to synthesize beta-carotene in the edible parts as well, fortifying them with the precursor to vitamin A and giving them a distinctive yellow hue. Call us when the rice also tastes like Skittles.

Don't be afraid to think outside the box: there's no reason a plant can't be pest-resistant *and* extradelicious. Genetically modified tomatoes that produce geraniol, a compound found in fruits and flowers that smells of roses, have a distinct flavor preferred in blind taste tests. GM onions that have their lachrymatory factor synthase genes shut down may taste slightly better, but they have another, more obvious advantage: without the enzyme produced by that gene, they don't cause your eyes to tear up when you chop them.

The GM food revolution isn't limited to simple vegetables. Even gourmet foods can be improved with biotechnology. The genome of the grape that gives us pinot noir has already been sequenced, and winegrowers are looking intently at the results, hoping to build grapes that are capable of producing award-winning wine in less than award-winning climate or soil. Some of these wineries possibly already use genetically modified yeasts that streamline the fermentation process and may reduce the formation of headache-causing compounds. Some are even field-testing disease-resistant grapevines. The coffee plant, too, can be improved in the usual ways. Pest-resistant plants already exist—but why stop there? If decaf (or, as we like to call it, "decoffeinated coffee") is your thing, reduction of caffeine content can be programmed directly into the genome of the plant. If you want your coffee beans unmolested by chemicals like methylene chloride, ethyl acetate, or supercritical carbon dioxide—all routinely used in various decaffeination processes—this research is for you.

Our sense of taste, complicated as it is by our sense of smell and the melange of compounds found in even the simplest foods, is difficult to break down at the molecular level. Genetic modification of food to improve flavor is going to be as much an art as it is a science, but it will, at least, make possible combinations you'd be hard-pressed to breed for. Eventually a trip to the grocery store could be an increasingly time-consuming experience. How does one choose between a strawboisenberry with the size and consistency of an apple, or a basket of strawberry-size boisenapples? The impact upon pie alone boggles the mind, and don't even get us started on smoothies.

A small red berry called "miracle fruit" suggests another tantalizing possibility. Eat one and miraculin, a compound naturally produced by the fruit, will bind to your taste buds, causing sour foods to taste sweet for up to an hour afterward. The effects of miraculin needn't be limited to the miracle fruit—if an existing edible plant could be genetically engineered to produce miraculin, its taste may be subtly or dramatically altered. Okay, so it won't really be a "miracle" until it makes English food tolerable, but, hey, it's a start.

Enough of the vegetarian fare; let's talk carnivore. In 2008, the U.S. Food and Drug Administration approved cloned meat for human consumption, although the current clones are far more likely to be used for breeding than for barbecue. It'll be a while before cloned animals make it onto anyone's plate, but a glass of milk from an animal with a clone somewhere in its family tree isn't far off. Eventually, we authors have hopes of a genetically modified turkey with six drumsticks and the addition of bacon-flavoring genes to . . . well, pretty much everything. This Thanksgiving, skip the tofurkey in favor of the genetically engineered turducken—one-third duck, one-third chicken, and one-third turkey. Three-thirds delicious.

Animals can certainly be tasty, but raising them in mass quantities has a significant environmental impact. Biotechnology is here to help. The Enviropig™ is a modified swine capable of digesting phosphorous and producing low-phosphorous manure that is less likely to encourage out-of-control algae growth in nearby groundwater.

Cattle are infamous for generating a great deal of methane as part of their digestive processes. That's a lot of greenhouse gas belched and farted into the atmosphere before the steak gets anywhere near the grill. Kangaroo flatulence, by comparison, has been shown (in experiments too horrible to contemplate) to contain hardly any methane. Researchers in Australia are trying to isolate the bacteria in a kangaroo's digestive system that allow it to perform this trick, to see if those bacteria can be transplanted into cattle.

Science isn't always pretty.

If cloned meat isn't your thing, how about vat-grown meat?

Animals take up a lot of space, and ranching them is expensive, morally contentious, and stinky. Instead of raising animals, you can raise animal cells in a bioreactor. Properly motivated, you can grow tissue today that . . . well, resembles meat. It's steak-like or poultry-esque but not quite ready for the grill, either economically or aesthetically. This technology still has plenty of room for improvement. Develop an inexpensive and palatable alternative to the slaughterhouse and you'll be doing the world, if not the dinner party, a favor.

Suggested projects:

- Carnifu—Soybeans that express various flavorful animal proteins. These can then be processed to create tofu with a distinctive meaty flavor. Nutritious, possibly edible, and extremely confusing.
- Cacaonut—Plants that produce cacao beans and also express various peanut genes. Get your chocolate in my peanut butter at the molecular level.
- Tomacco—Homer Simpson accidentally created the to-macco by fertilizing his tomato and tobacco fields with plutonium. A genetic tomato/tobacco hybrid could have the addictive properties of nicotine, the nutritional value of half a tomato, and the taste of a used ashtray. Get your kids hooked on veggies. Literally.

MEDICINE

Your immune system is engaged in a lifelong struggle against a host of invaders, but biotechnology is here to help. While there are many bacteria with which we have mutual nonaggression treaties and even depend upon (to aid digestion, for example), there are others that would be happy to hold a toga party in your appendix, given the chance. Antibiotic compounds can fight many infections, but they aren't a panacea; bacteria can develop resistances to these compounds through repeated exposure. In the race between bacterial evolution toward antibiotic resistance and the development of new antibiotics, human triumph is hardly assured.

Many traditional pharmaceuticals were originally derived from plants and animals, but getting them out of that plant or animal can be a chore. Genetic engineering can be used to transfer a useful gene from a difficult organism into one more pleasant to work with. Rice can be more than pest-resistant and fortified: it can also be engineered to express a cholera vaccine primed for oral administration, swapping an injection for a snack.

Tobacco is another widely farmed product with biological versatility. Gene knockouts can reduce the amount of certain carcinogens in the smokable leaves, and it has recently become possible to produce large amounts of certain human proteins in the tobacco plant. Since tobacco is easy to grow and produces lots of biomass, the plant is an ideal protein factory. Getting the proteins out of the plant while they are still functional can be tricky, but worth the effort.

Lucky for humans, some animals can be useful to medicine as more than mere test subjects. Scientists have been producing vaccines in chicken eggs for more than fifty years, but only recently have genetically modified hens been capable of producing human proteins that can easily be harvested, sunny-side up. The end goal of medicine is typically human therapy, but in the future it's likely that our medicines will be produced by lower life-forms.

Suggested projects:
- Cacaojuana—Cacao beans that produce THC, the psychoactive ingredient in cannabis. Righteous brownies, dude. (For medicinal use only, of course.)
- Cherries that produce a nasal decongestant—Finally, a cherry-flavored cold medicine that actually tastes like cherries.
- Juniper berries that produce quinine—Gin and tonic flavor in a single plant.

PETS

Anyone who has lost a beloved family pet understands the desire to somehow resurrect Fido or Mittens. How many replacement pets are there with names that end with ". . . , the Second"? If a look-alike just won't do, cloning is an option. Cats are already being commercially cloned, though the fees are so gargantuan that a trip to the pound for something merely Fluffy-esque is by far more practical.

If money is no object, though, why stop at cloning? If you can be bioenhanced, why not your animal companion? With a little ingenuity, most of the enhancements in the previous chapter can be applied to pets. A myostatin-inhibited Chihuahua could put on enough muscle to make a pit bull think twice; and a cat with a longer life span would be able to demonstrate its complete disdain for your existence several years longer than a regular cat has the energy to. If your cat's dander has been genetically altered to be nonallergenic, so much the better.

If you're more interested in aesthetics and your pets are likely to suffer the indignities of "precious" little outfits or outrageously dyed fur, perhaps you'd be interested in a fluorescent pet? Genes for fluorescent jellyfish proteins have been successfully inserted into mice, rats, pigs, and kittens. Fish can also be modified in this way, and they typically live in tanks with a UV light that will cause the protein to glow brightly.

Biotechnology can even produce the petless pet. Pet rats are messy and hell on furniture. If you're looking for a rat that won't get on your nerves or interfere with your pathological neatness, take a computer chip with electrodes on its surface and slap a few tens of thousands of rat neurons on it. Hook the chip up to a computer, and you can "teach" the cells on the chip to control a virtual rat or play a flight simulator—it's been done, and don't think we're going to stop there. Why not combine the neurons of a biological pet with the convenience of a virtual pet? The Tamagotchi™ has nothing on VirtuaRat, your friendly pal with all of the brains and none of the feces. When you're not slipping him a bit of VirtuaCheese (available for download from our website at a modest fee), you can teach the same chip to gold farm for you on the massively multiplayer game of your choice. Why not outsource a repetitive task when you can *downsource* it to a lower life-form? Just don't give it unlimited access to the Internet. Besides the global thermonuclear war scenario, no intelligence, however meager, should spend too much time online.

Suggested projects:

- Intelligent pets—A goldfish is okay, but one that does your taxes is something special.
- "Genetic leash"—Ever have a dog that got lost or ran away? A little psychological tweaking could ensure your pet's return at the end of each day. No fences or leashes required.
- "Pocket pets"—Some animals, like panda bears, would be fun to keep at home, but they're simply too cumbersome to care for. Fiddle with the natural course of development, however, and your next pet could be a miniaturized elephant.

CHIMERAS AND OTHER HYBRIDS

Why have one pet when you can have two? No, scratch that. Why have two pets when you can have a single chimeric one? An eagle would be cool, and a lion would be even better, but a griffin would be *sweeeeet*. Just sitting around hoping for some hot bird-on-cat action isn't going to get it done, though. If you want to make a chimera, take a lesson from nature.

Marmosets, for example, are so frequently chimeric that they often have children on behalf of a sibling, because some of their sperm or eggs originate from cells they picked up from a brother or sister in the womb. Mixing cells from one species with cells from another is more problematic, except under very special circumstances. It's much easier to produce a mouse with a few canine immune cells than it is to make a typical mythological chimera; say, a creature with the body of a mouse and the head of a pit bull.

Genetic manipulation can be used to introduce one animal's genes into another, but it's unlikely that the insertion of a single chameleon gene into a mouse is going to produce the world's first ninja rodent (oh, how we've *tried*). Creating such obvious hybrids will not be a simple task, but the possibilities are far too interesting to ignore.

Assuming for a brief, shining moment that the science and expertise to create these freaks of nature comes about, where do you begin? We suggest a mythologically tested combination. While the following creatures have never existed, they're familiar enough to most people from folklore that you might be able to get away with one or two before the villagers show up at your door with torches. A classically trained genetic engineer is a force to be reckoned with.

- Cockatrice—The body of a rooster with the tail of a serpent. Makes a platypus look relatively normal.
- Jackelope—A jackrabbit with antelope horns. You'd be a fool not to.
- Pegasus—A horse with wings. Beats a hybrid for mileage, too.

We're going to avoid the animal/human mixes and suggest you also take a pass on them. In addition to the technical issues, it's also seriously uncool. Don't set out to build a creature that is morphologically unique yet essentially human, dooming it to a lifetime of freakish solitude and physical incompatibility with the rest of humanity. Even if you commit yourself to producing a community of like-bodied individuals, how well do you think they'll fare as the planet's first man-made minority group?

- Cat-woman—A woman with feline features. Doomed to a life of marriage proposals from anime geeks.
- Centaur—A man's torso attached to the neck of a horse. Sure, it sounds like fun, ladies, but consider the logistics.
- Harpy—Half woman, half bird. Wings are cool and all, but you don't necessarily want to trade in your hands for them.
- Manticore—The body of a lion, the head of a man, and the tail of a scorpion. What, lions aren't dangerous enough for you already?
- Medusa—A woman with snakes for hair. If you want to create something famous with bad hair, try cloning pretty much any 1980s celebrity instead.
- Minotaur—In Greek mythology, this offspring of forbidden lust is a man with the head of a bull. The moral here is: some lusts are forbidden for a reason.
- Satyr—Half man, half goat, all randy.
- Sphinx—The body of a lion with the head of a human. Big on knock-knock jokes.
- Spider-man—Major copyright issues, and in this case, the webbing comes out of his butt. Remember, with great power comes great responsibility.

Science is about embarrassing those fools at the Institute who said you were mad, not embarrassing yourself. Please, avoid lame mixes such as:

- Tuba—The head of a goat and the body of a snail. When you need an animal that will eat garbage *very slowly*.
- Unicorn—Too obvious. Besides, having a unicorn around is going to open you up to an onslaught of virgin jokes.
- Hippogriff—Half griffin, half filly. A griffin is already half lion, half eagle. You go too far.
- Grape Ape—Half ape, half grape. The wine is *terrible*.
- Winged monkeys—Giving wings to an animal known for flinging its own excrement is neither the height of scientific inquiry nor wise.

Not all chimeras are winners, but your experimental rejects could easily find new life in the discount video bin. For example, "frog" plus "giant snake" equals "frogaconda." And as any fan of *Jurassic Park* can tell you, even extinction can't completely rule out experimentation. The Pyrenean ibex (a wild mountain goat) was declared extinct in 2000. A clone was generated from a frozen Pyrenean ibex skin sample and brought to term in a domestic goat (though the cloned animal died shortly after birth). A gene recovered from the remains of an extinct Tasmanian tiger has been inserted into mice, giving scientists an opportunity to observe its biological function. Granted, the Tasmanian tiger died out in the 1930s, not millions of years ago, so dinosaurs are likely to be a stretch—but it's worth a shot. Seriously, what could go wrong?

Suggested projects:
- Octocroc
- Velocisquid
- Tarantapotamus
- Scorpadillo
- Great White Bee
- Piranhabat
- Giraffeosaurus

CONCLUSION: THE GRASS IS ALWAYS GREENER ON A BIOENGINEERED LAWN

Unless, of course, the lawn has been bioengineered to be purple. Microorganisms, plants, and animals are a (relatively) safe, (tentatively) legal, and (debatably) moral alternative to human experimentation. A well-chosen initial research program using lower life-forms will cultivate good lab skills, hone your mind, and support its own costs. Every beginner makes mistakes; better they lead to a spoiled Petri dish than to that-which-humanity-was-not-meant-to-know.

Many find working with lower life-forms to be fulfilling, and devote their lives to a better understanding of or the manipulation of a single organism. Others work with lower life-forms only as a prelude to the brass ring of biological manipulation—humankind. Eventually a handful of these researchers will reach the pinnacles of brilliant and ballsy, and we'll all be living in a world where the human genome is in a permanent state of "under construction."

Tread carefully. Never forget that genetic tinkering has as

much potential to harm as it does to help. Dr. "Hawkeye" Pierce we'd trust with our genes; Dr. Moreau . . . not so much. Once you've mastered genius sharks and killer tomatoes, your greatest challenge—human cloning—could easily lead to your downfall.

SIX

HOW TO DEFEAT YOUR OWN CLONE

(and Look Good Doing It)

It would take three clones to beat the original Homer—I mean four! . . . suckers. —HOMER SIMPSON, *THE SIMPSONS*

You are your own worst enemy. —UNKNOWN

The biotech revolution holds both wonders and perils alike, the outcomes of which are driven mostly by our choices as individuals and as a society. When technology is applied with imagination and wisdom, it enriches both our material and intellectual lives. When either is lacking, there's a good chance that impetuous implementation will come back to bite us in the ass. Perhaps the most certain thing that can be said about the near future is that it will be filled with people, some of whom are sure to exhibit both a lack of imagination and wisdom. We hope you won't be one of them—or two of them.

If you are cloned, or clone yourself for the wrong reasons, you may end up with nothing more than a cunning adversary, and a contest between yourself and your clone won't necessarily be a coin flip. A realistic clone will be younger than you are—generally an advantage in a fight unless the clone is still in kindergarten—and will benefit from advances in modern medicine and nutrition, possibly including significant bioenhancements. A clone devoted to the study of unarmed combat or biotechnology law is probably going to have the upper hand over an original devoted to fantasy football or who owns every season of *Law and Order* on DVD. Genetic similarity is no guarantee of a level playing field.

In previous chapters we strayed only occasionally into the land of speculation, and concerned ourselves mostly with scenarios that seem likely to occur in the near future. The clones that appear in popular culture, however, tend to be improbable or inexplicable, but not quite impossible. Arthur C. Clarke once noted, "If an elderly but distinguished scientist says that something is possible, he is almost certainly right; but if he says that it is impossible, he is very probably wrong." Aiming to be comprehensive, we'll take this opportunity to suggest strategies for dealing not only with the sorts of clones created by biologists, but those created by a scriptwriter, too. Chances are you'll never have to deal with a recalcitrant clone that mysteriously shares your memories, but if you do, at least now you'll be prepared for it.

WE HAVE MET THE ENEMY AND HE IS US

The degree to which your clone poses an immediate and personal threat depends upon how and why the clone was created in the first place. Your clone will have its own point of view, and understanding it is critical if you want to achieve either peace or victory.

Basic genetic clone

This clone is genetically identical to you, but nothing more. Starting as an embryo with a copy of your (formerly) unique genome, this duplicate gets a fresh shot at life. Probability of encounter: plausible. While human reproductive cloning is still just theoretical (not to mention illegal), a basic genetic clone will be the first type produced, if and when human cloning finally happens. Average threat level: nil. If you can't defeat a baby, your problems are beyond the scope of this book.

"Age-matched" clone

This clone has undergone various procedures to make her close or identical to you in "age" (either perceived or actual). In contrast to a basic infant clone, this version looks much more like the original thanks to a parallel biological development. Probability of encounter: unlikely. If manipulating the aging process were easy, companies like Revlon and L'Oréal wouldn't exist, but it isn't, and they do. If you encounter an age-matched clone within your lifetime, its creator deserves a Nobel Prize. Average threat level: low. This type of clone is similar to an identical twin—she looks like you, but shares none of your memories or experiences. An age-matched clone may be difficult to identify at first glance, but a quick interview by friends or family should do the trick, especially if the clone hasn't had the benefit of some sort of accelerated education (the baby talk and thumb sucking are usually a dead giveaway).

"Mind-clone"

This clone is one of the most common types portrayed in pop culture. A mind-clone shares all your memories up to the point of cloning, and may even believe that it is the original version. Producing this type of duplicate will require much more than genetic manipulation, but may eventually become possible with advances in neurology and computer science. Probability of encounter: don't hold your breath. This is the sort of thing that makes for strong ticket sales at the box office, but requires a level of technology far beyond the foreseeable future. Still, fifty years ago, not many people thought we'd be solving crimes with DNA "fingerprints" or cloning sheep for shits and giggles. We live in an age of rapid biological innovation, and you can never be too safe. A more plausible variation on this theme is a clone that has been trained to mimic you and is educated thoroughly about your past in lieu of any actual memory transfer. Average threat level: medium. A foe that knows your darkest secrets is a foe to be reckoned with, indeed. Thankfully, in this case, the reverse is also true.

Note: In pop culture, mind-clones are almost always depicted as age-matched as well; otherwise, this is just a very smart baby that shares your fear of mimes. Many (but not all) of the clone-defeating techniques useful against an age-matched clone are also applicable to a mind-clone who resembles you. In terms of combat, we'll treat the mind-clone as a complex variant of the age-matched clone.

Bioenhanced or modified clone

This clone is based on your genetic template but has been genetically modified to be faster, smarter, stronger, etc. Probability of encounter: it could happen. Cloned animals have already been created with genetic modifications, so it shouldn't take much to repeat the process in humans. Granted, the first genetic upgrades will probably be stuff like abnormally white teeth and resistance to the flu, but when human clones start showing up with shark incisors and a resistance to death . . . watch out. Average threat level: high. Defeating your identical twin is no picnic; defeating one that's engineered to be better than you at everything isn't even a bag of trail mix with all of the M&Ms picked out.

Combination clone

This clone falls into multiple categories from those listed above. Potentially, you could have a mental and physical doppelgänger that differs from you only in her overwhelming superiority. Probability of encounter: doubtful. You'll likely never have to deal with an age-matched, genetically enhanced clone who also shares your complex neural makeup, but if you do (and this is a *big* if), average threat level: sucks to be you.

Other

Identical alter-egos may also arise from several other science fictiony-type phenomena—transdimensional double, time-travel duplicate, teleportation mishap, etc. Probability of encounter: hard to say. Biological cloning already exists in various forms, but these other clones are exercises in hypothetical quantum physics. We can't say these clones are likely, but with all the stuff that may or may not go down at Area 51, we can't say they're impossible either. Average threat level: variable. These "clones" are all potentially dangerous, but different rules of engagement may apply, depending on the circumstances.

The basic genetic clone will certainly be the first type that we encounter as the revolution unfolds, but the others make for more interesting (and perilous) combat scenarios. Fear not—we have you covered on all fronts.

ORIGINALS ARE FROM MARS, THEIR CLONES ARE FROM VENUS

Your relationship with your clone will influence (or determine) any conflict that arises between you. Are you parent and child, tyrant and minion, or aging fascist and unwitting organ donor? Did you willingly supply your DNA, or was it stolen from you with malicious intent? Clones, like other human beings, are most likely to act as adversaries when you've given them a good reason to fear or dislike you, but even the purest of intentions are no guarantee that you and your clone are going to get along.

If you decide to clone yourself, barring artificial aging of the clone or a stint in the cryogenic freezer for you, the clone will be born when you are in adulthood. This clone will be, for all intents and purposes, your child. Like other children, your clone will eventually become a teenager, and your mother will smile contentedly knowing that her ages-old curse, "I hope that when you have children, you have one exactly like you," has come to pass. While you may share a genetic identity, you will have distinct upbringings and a generation gap between you. Good parenting and an early realization that your clone is an individual is your best bet—it works for (most) ordinary families. If, instead, you treat that same clone as a convenient source of medical material, you'll have an enemy for life—your very existence is a constant threat to a clone earmarked for parts.

A clone somehow aged to match you may, with careful work, resemble you physically but differ significantly in personality. Such a clone can be a competitor, a compatriot, or a mere acquaintance. Effectively the two of you will be twins separated at birth, and your relationship will depend on your dispositions. You and your engineered sibling could be inseparable, or you could be famous for your annual Thanksgiving shouting match. Failure to accept a similarly aged clone as a brother or sister is no different from turning your back on a long-lost twin. If you were unwittingly duplicated, there's no guarantee that your clone—young or old—will want to be part of your life once one of you discovers the other.

A clone who shares your genes, age, *and* memories is not

going to be satisfied with a sibling relationship, however con-
genial. As far as the clone is concerned, he or she is *you*. Imagine going to work or home to your family to find an identical
you at your desk or playing with your children. If offered convincing proof that you are not the original, would that make
your feelings of accomplishment in your career or love for
"your" offspring any less real? It's hard to imagine how such a
clone would be anything but desperate to live as you do—
unless, of course, you hate your life. If so, your clone will be
happy to be relieved of the burden that is your daily grind, and
the worst you two will have to contend with is your seething
jealousy of another you out there somewhere who might not be
as miserable as you are. Frankly, if you would be willing to do
battle with a clone to see which of you would get stuck living
your life, cloning is the least of your worries.

There are few reasons why anyone would create a clone that
shares your memories, and most of them ride the bad-idea train
to the end of the line. An exact duplicate would be useless to
any nefarious plot to replace you because it would, in effect, replace you with . . . another you. Throw in some *Manchurian
Candidate* mind control and maybe you're approaching a cunning scheme, but it seems to us that unless you're a president,
dictator, or Fortune 500 CEO, there's little chance anyone
would go to the bother and expense. If you wish there were
more hours in the day and have considered duplicating yourself,
imagine a conversation wherein you attempt to convince your
duplicate that, while you're out partying, said duplicate should
be standing in line at the DMV. If you think divvying up chores
between roommates is tough, try dividing a single life equitably

between two (or more) identical human beings. Whether your clone wants to take over your life or merely call dibs on the best bits of it, having a genetic and experiential duplicate around is a recipe for discord.

SO IT'S COME TO THIS: YOU HATE YOUR OWN GUTS

It's entirely possible that you and your clone will cast off the public misconceptions and become super best friends forever, but then again, maybe not. There are plenty of reasons to be at odds with your genetic double: one of you wants to steal the other's liver, for example. And then it's *on*.

The authors are neither great military tacticians nor experienced fighters, but in a clone war we do have a few points in our favor besides our academic work in biotechnology. Most rivalries between you and your clone can be considered an unusual case of sibling rivalry, and we both have multiple siblings. Additionally, as fundamentally unserious people, we've wasted a great deal of our lives consuming media in which various protagonists deal with their arch-nemeses. Think of us, your clone-defeating gurus, as a hybrid of Bill Nye the Science Guy and Michael Bay.

One final thought before we prepare you for a possible doppelgänger showdown: Always remember that your clone will be just as much a human being as you are. Would you sucker punch your mom for making you eat your vegetables? Why then resort to violence if your clone is merely being a bit of a pest? There are many ways to contend with your troublesome genetic double. Physical confrontation should be a last resort.

THE RULES OF ENGAGEMENT

Dealing with a clone is a lot like dealing with any other bother-some individual, except that, in this case, the individual shares your genes (and possibly more). If you've decided that your clone must go, we suggest the following general strategy:

Step 1: Know your enemy

First off, where did the clone come from? If you cloned your-self, but now you're regretting it, maybe you should think twice before taking it out on the clone. If instead you were cloned by someone else as part of some grandiose scheme, then your clone might not even be the real enemy. Second, what kind of clone is it? Outsmarting a neurally enhanced genius will be a lot harder than outwitting a grade-school version of yourself. Last, what is the conflict at hand? If your clone is trying to assassi-nate you and steal your identity, you're pretty much entitled to use any means necessary; if he's merely trying to steal your date for Saturday night, you might want to think twice about that grenade launcher.

Step 2: Analyze your strengths and weaknesses

If you believe that your clone is exactly like you, then you haven't been paying attention. You and your genetic double may be similar in many ways, but you both exist as individuals, and that alone makes you different. The question is: What do you have that your clone doesn't (and vice versa)? Your greatest chance of success is to play to your own strengths and to your clone's weaknesses.

Step 3: Plan your strategy

Will you choose to outsmart your clone or outmuscle him? If your clone is trying to charm your friends and family, why not beat him at his own game? There are many battlefields upon which your nemesis can be defeated: social, intellectual, physical, or genetic. Personally, the authors would prefer to defeat our clones at some sort of roundtable political debate. Just seems classier.

Step 4: Act and adapt

Be prepared to think on the fly. Plans rarely unfold as anticipated, and it's always best to have a backup strategy. For instance, if your attempt to frame your clone for insurance fraud now has you running from authorities instead, you might want to look into obtaining some legal counsel, or at the very least, a convincing fake mustache. Your original strategy may also bring up unexpected issues; perhaps you're planning to jump your clone in a dark alley, but from a psychological standpoint, it might not be so easy to kick yourself in the face.

Every clone will be different, and every situation will require a unique resolution—there is no single clone-defeating strategy, just as there is no single people-defeating strategy. We can't tell you exactly what you'll need to deal with your specific cloning predicament; however, we can supply you with a host of potential scenarios designed to prepare your mind for the various types of clones you might have to defeat. We'll start with the basics (the most plausible and simplest to defeat clone antagonists) and work our way up to "Tinseltown madness."

YOU'VE GOT A REAL ATTITUDE PROBLEM, MISTER

As the original copy, you may find yourself looking down upon your "lesser genetic replica." This is a trap. As far as nature is concerned, you two are equal, and you're not going to win this battle just because you've got a superiority complex. Besides, anyone who's seen *Rocky III* knows that when you get cocky, you get punched in the face by Mr. T. It's just that simple.

Defeating your own clone will require humility above all else. By now you should understand what you're up against, so don't underestimate yourself.

HOW TO DEFEAT A BASIC GENETIC CLONE

The simplest clone imaginable is a newborn spawned from a copy of your unique genome. Why you would want to "defeat" such a (presumably) cute baby clone is beyond us, and we're not going to help. Figure this one out for yourself. Or better yet, don't.

Fast-forward from infancy to rebellion. Your harmless infant double is now a whiny teenage punk and you're starting to realize what a miracle it is that your parents never bludgeoned you with your own Nintendo GameCube during your rebellious high school years (we've asked ours—there were a lot of close calls). Adolescent defiance is a normal human emotion and a critical part of developing one's individuality. Given that your clone is even more similar to you than you are to your parents, the chance for rebellion is high, especially since your face will serve as a looking glass into your clone's future. Imagine taking orders from an older, fatter, balder version of yourself. You'd be a little uppity too, wouldn't you?

Yet the conflict between you and your clone won't necessarily be grounded in prepubescent hormones. For most people, there is a basic human need to define themselves as individuals, and having to live with a genetic copy of yourself (particularly one that acts like a parallel version of you on a slightly altered timeline) could throw a serious wrench into the works. As the first human clones arrive, there will undoubtedly be issues to deal with, particularly if the originals are still around. But matters like "Which clone is the better bass player?" or "Who gets the top bunk tonight?" are relatively petty and can be worked through rationally. At worst, you and your clone may need to simply part ways and never see each other again.

Rather, the biggest danger of having a clone (as far as Hollywood is concerned anyway) is that he's in a unique position to steal your friends, your belongings, or even your entire life. Today, your identity can be stolen by an anonymous hacker with your social security and credit card numbers; in the future, it could be stolen by a guy with your blood type and face. Of course, if the clone is twenty years younger than you, impersonation will obviously be much more difficult, which makes identity theft unlikely, but not impossible. Imagine a situation where someone—the government, your employer, your spouse—

produces a younger, healthier version of you because the original is no longer up to par. Everyone loves the new "Stan, version 2.0," except that *you're* Stan, and you're finding yourself seriously outclassed.

While it's tempting to go around defeating everyone who's younger and more likable than you, modern society just doesn't work like that. In fact, that's a good way to get yourself arrested. A basic genetic clone is simply a person who shares your genes, and any apparent attempt by that clone to "take over" your life is most likely a delusion caused by your own personal insecurities. So unless your clone is out smuggling plutonium, organizing cockfights, or running for public office, you should probably just let him live his own life. However, if your clone has been artificially aged to match you in appearance, the boundaries of "his" life and "your" life may begin to blur. And that's where things get interesting.

HOW TO DEFEAT AN AGE/MIND-MATCHED CLONE

A clone that has been artificially aged to resemble you is a relatively convincing identity replacement. If that clone also shares your memories, you've got serious trouble. Realistically, if your clone shares both your age and memory, the duplicate won't be preferable to the original in any way, but in movies, books, and comics such clones are often considered expendable just so long as one copy of the character survives. No one seems to care which one lives and which one dies (except the clone and the original), despite the fact that both are living, breathing human beings. Once your clone wakes up and starts experiencing life, the two of you are distinct individuals.

Ideally, your clone can find her way in the world without trying to steal yours, but if rotten luck or your meddling impedes your clone's progress, the following guidelines will help you cope with an age-matched (and possibly mind-matched) duplicate that poses an immediate and serious threat to your identity.

Utilize your personal belongings

As long as your clone hasn't infiltrated your life yet, you should still have possession of all your stuff, which is an advantage on multiple levels. First, it provides you with numerous opportunities that your clone won't have access to. Second, it makes it easier for others to spot the faker. Here's a few examples to get you started:

- Keys—Provide access to your house, work, vehicles, safe-deposit boxes, etc. If your clone is attempting to supplant you, the simplest thing to do is lock her out. Life won't be easy for your genetic double without a roof to sleep under or a car to get around in, and she probably won't be fooling anyone if she can't even get in to "her" own gym locker.

- Money—Cash, credit cards, bank account info, etc. Donald Trump's mind-clone might remember what it was like to be rich, but simply remembering won't get him a penthouse suite. Keep a close eye on your finances; your clone may be so busy working double shifts at the 7-Eleven that she won't have the time or resources to even think about replacing you.

- Identification—Driver's license, passport, social security card, birth certificate, etc. Most likely, your clone will be entitled to her own ID cards so that she can be a productive member of society, but without *your* ID, she'll have a difficult time impersonating you at work or even at your local bar. If your best friend is suddenly known around town as "McLovin," be wary of an impostor.
- Clothing—For some people, their fashion sense is as unique as their genomes. Your attire can be duplicated, but probably not quickly or easily, and your clone is sure to rouse suspicion if she's sporting an entirely new wardrobe.

Remember, your belongings not only distinguish you from your clone, but some of them may be useful in "combat," like those ninja throwing stars you ordered online (or a copy of this book). Anything that you have and your clone doesn't is an advantage (excepting an outstanding warrant or a tapeworm).

Secure your identity

Quite possibly the easiest way to get yourself replaced by your clone is to keep the situation a secret. If no one knows your clone exists, then who will ever notice when you've been replaced? Your best strategy is to quickly explain the situation to the necessary parties and convince them to help you.

What if your clone gets to your companions first? If you notice friends or family acting strangely, your double may have already taken steps to supplant you, and you'll need to proceed with caution. In this case, it's likely that your duplicate has posed as the original and did exactly what you would have done: warned everyone to be on the lookout for impostors. Your acquaintances will be understandably wary of you or anyone who *appears* to be you, but all is not lost. Your best bet is to request an opportunity to prove your identity. Be careful—if your colleagues are sufficiently spooked, they may shoot first and ask questions later.

Once your chosen cohort has agreed to hear your side of the story, act decisively and be prepared: you may only get one attempt at this. If your friends are wise, they'll be analyzing you closely, waiting for a slipup. A forgotten birthday or a suspicious freckle could mean the difference between success and failure. You may need an exit strategy if the interview starts heading downhill; pepper spray and a pair of quick running shoes should be your accessories of choice.

Once you've reestablished your identity, you'll still need to prevent future attempts at misdirection from a hostile clone. Keep a list of unique features such as birthmarks, scars, and childhood memories handy, and distribute it to those you trust, because the last thing you want is to be mistaken for your own double after you've already convinced everyone that he's not to be trusted. Also, consider establishing code words or a secret handshake that could be used for easy reidentification in the event of a mix-up.

It's also a good idea to start thinking defensively. If you have the means, try upgrading your locks to retinal or fingerprint scanners that your clone can't easily circumvent, or at least get rid of those spare keys—with a doppelgänger on the loose, it's way too easy for them to end up in the wrong hands. The more roadblocks you erect, the harder it will be for your clone to infiltrate your life.

The lesson is: work quickly to defend your life so that your clone is left on the outside looking in. The faster you secure your identity and your belongings, the harder it will be for your clone to steal them.

If you're dealing with an age-matched mind-clone, your doppelgänger will almost certainly try to infiltrate your life, because really, it's her life, too. She knows your email passwords, PIN, and favorite dirty limerick just as well as you do. If the clone has also been competently engineered to share any distinguishing marks, confirming your identity may prove difficult or impossible. In fact, you very well may *be* the clone. The best you can hope for is to persuade your friends and family to enter into a temporary truce while you and your duplicate sort this out for yourselves.

Experience is your ally

When dealing with an age-matched clone, take advantage of your life experience, particularly if his "maturity" is only skin-deep. For example, advanced degrees in math and psychology would help you best your clone in a winner-take-all game of Texas Hold 'em. Or, if you've spent the last seventeen years running black ops in Sudan while your clone was busy trying to relive your privileged childhood in the Hamptons, you probably have a leg up in the combat department. Realistically, your edge in experience is likely to be more subtle than a Ph.D. or sniper training. For instance, if you were flighty and impulsive when you were younger but mastered self-discipline as you matured, your clone may have the same tendencies. Make the most of your newfound inner tranquility while you wait for your clone to make a foolish mistake.

Experience is also particularly useful in social situations. For example, every family has a collection of subjects best not discussed at the dinner table, lest a shouting match begin (religion, politics, Notre Dame football, etc.), and your clone isn't necessarily in on the cease-fire agreement. On the other hand, he doesn't have the associated lifetime of resentment and psychological baggage, so maybe it's a draw. Your parents might actually enjoy having a clone of you around who isn't constantly bringing up the skeletons in the family closet. Try to get back on your family's good side while sending your clone down several hilarious dead ends ("So 'Jim 2,' what's your opinion on gay marriage?").

If family dinner with your clone is a minefield, dating is a minefield on fire. Your clone may find himself attracted to the same kind of women that you're into—if so, keep in mind that he has yet to make every relationship mistake you've ever made, and is likely to be prone to more than a few. A clone can easily be misled into habits you had a hard time breaking, especially if he's unaware of his predisposition to them.

Even experience with your own body is an advantage over your clone, because you know and understand all of your own personal weaknesses way better than he does. This can be particularly useful if your clone has been age-matched, but only just popped out of the tank last week. You've had decades to figure out exactly how long you can stay out in the sun without risking a painful sunburn—your clone hasn't.

In a struggle with your comparative equal, use your experience and self-knowledge to defeat yourself. Zen, huh?

Experiences like those above provide clear advantages in combat, but what if they could be faked by your clone? "Memory" and "experience" aren't exactly the same thing, but for experiences that are primarily intellectual, remembering the event is the most important part. For example, you may have spent years in law school studying civil litigation or international property rights, but your mind-clone won't need all that time hitting the textbooks; the *memory* of the education will be just as good as the education itself. Conversely, what use is your fancy-pants degree if you can't remember any of it?

Fortunately, if you're dealing with a mind-clone who shares your memories, you may still have the advantage of a lifetime of physical conditioning. For example, just because you understand *how* to balance on one foot doesn't mean you can actually *do* it without practicing first. Now extrapolate this to a much more complicated physical activity like a martial art. Training your body and peripheral nervous system is an arduous process in which practice and repetition are key. If you happen to be a black belt in tae kwon do, your untrained clone is in a lot of trouble. Even if your clone "remembers" all those years of training, she won't have the associated muscle memory that went along with it, and her attempt at a tricky flying roundhouse kick is more likely to involve a strained hammy and a butt land-ing than a successful strike. Remember, even Baryshnikov didn't pirouette his way out of the womb—he gracelessly slid into existence just like everyone else. Muscle memory is some-thing that takes years to fully develop, and while you've had your whole life to prepare for this, your clone may have had only a few months or less.

Time is on your side . . . for now

If there's one advantage that you will always have over your clone, it's time. Barring a few implausible sci-fi, time-manipulation sce-narios, your clone will never come into existence before you do. This means that you always get the upper hand in terms of preparation and training—if you're not too lazy, that is. Keep in mind that if you've spent your life with a Twinkie in one hand and a Coors Light in the other, you're probably not at your ideal fightin' weight. The authors, for example, were raised in the Midwest on Mom's home cooking, but our clones might be raised in a vat of growth hormones and steroids. Heed our ad-vice: Slim-Fast and Pilates work wonders.

Time also provides you with an opportunity to study up on and outsmart your clone. Reading this book is a strong start, but don't throw away your library card just yet; a little extra knowledge of lock picking or wilderness survival techniques never hurt anyone. Every new skill or tidbit of information can be used in your fight against your mischievous double.

Be careful, though, because time can also work against you. You may have the disadvantage of a body with years of wear and tear, while your clone is just hitting his prime. It's not going to be easy to defend yourself in a fistfight if that old rugby injury keeps acting up.

The lesson here is to train yourself constantly and keep your body healthy and fit for as long as you can. Procrastination may be your biggest enemy—start preparing now.

Legal options

Sure, your clone shares your DNA, but does that mean he has any rights? This all depends on the current political climate at the time of your duplication. In a "clone hostile" system, there's a chance that your clone is entitled to absolutely nothing. There's an even better chance that he's entitled to a subpoena, a restraining order, and a court-ordered GPS tracking bracelet. We're not saying such poor treatment of clones is right (in fact, we've repeatedly said the opposite). If your back's against the wall, however, you might be forced to take advantage of an unfair legal system.

Even if you think you have the upper hand with respect to cloning law, don't go running off to your lawyer buddy just yet. If your clone commits a crime while looking like you and leaving behind DNA evidence that suspiciously resembles your own, you may have a serious problem. Even if you're the one trying to pin the crime on your clone, the authorities still need to be able to properly identify the two of you. Who are the cops going to believe: you, or the guy claiming to be you?

If you created the clone illegally, no subterfuge on her part is necessary—the clone is walking proof of your complicity in an unlawful act. Such a clone can hold his or her own creation against you, as blackmail or evidence in a court of law. The authors are not qualified to write *How to Defeat Your Own Clone's Lawyer*; the best we can do is warn you not to do anything you wouldn't want to admit before a jury. The same advice applies if you're interested in cloning someone else, only more so. If you fail to get his or her consent, you'll get what's coming to you: the clone teaming up with the original to take you to prison, the poorhouse, or both.

As risky as an appeal to the law can be, it's far more dangerous to involve those who operate outside the law. Hire an assassin to take out your clone and you may end up buying yourself a bullet. If you manage to get on the wrong side of organized crime, a cabal of drug lords, or the Girl Scouts, you might be able to maneuver them into taking out their ire on your clone—after which you should be prepared to live out the rest of your days posing quietly as a Guatemalan cattle rancher.

Battle of wits

An age-matched clone may look like you, but depending on the method used, much of this aging could be superficial. For example, accelerated aging might speed up your clone's baldness and crow's-feet, but it won't replicate your torn ACL or your smoking-induced emphysema. While your artificially aged double may boast superior physical prowess, your years of worldly experience should at least give you an advantage on the mental playing field.

In a battle of wits with your clone, wisdom and self-understanding are your closest allies. If your particular genetic disposition makes you a very happy drunk, for example, you could try reasoning with your clone over drinks; your intoxicated genetic double might even call the whole thing off and buy you a pint. If you're renowned for your gullibility, perhaps your clone is as well—you can play the old "you've got the wrong guy" prank, with a twist; convince your clone that you're not the original, but instead another perfectly innocent clone. Or just trick the clone into joining one of those religious cults while still (mentally) young and impressionable. Whatever is easiest.

However, if you're dealing with a mind-clone who shares all of your memories, a battle of wits may be relatively pointless, unless of course you plan to cheat. Except that if you're the cheating type, then maybe so is . . . well, you get the idea.

Finally, you may be tempted to taunt your genetic double with embarrassing personal information, enough to infuriate him into making a tactical error. Resist the temptation—mocking your opponent usually works best when your opponent is not you. Ridiculing an exact copy of yourself will almost certainly lead to the "I know you are, but what am I?" defense. This is asinine at best.

Physical combat

If a fight with your clone is completely unavoidable, you should at least be ready. Remember how your parents always told you "Life isn't fair"? Well, neither is clonal combat, though that doesn't mean it can't be unfair in your favor. Here are a few tips to get you started:

- Fight dirty. Hair pulling, eye gouging, ear biting, crotch kicking, and sand in the face—all these techniques are unexpected and extremely effective. Plus, you don't need years of ninja training to pull them off.

- Don't take a knife to a gunfight. Admit it, you've always wanted those nunchakus but you could never justify the purchase. Well, now's your chance. You want that cane with the sword hidden inside? Go ahead. How about that ridiculous-looking trident? Tasers, sniper rifles, Gatling guns, trebuchets—it's all fair game now. While your clone spends all his time training for a fistfight, he's gonna look pretty stupid when you show up in your amphibious assault vehicle. Actually, you'll both look stupid, but at least you'll be the one driving a floating tank.

- Seize the moment. If you're going to do battle with your clone, there's no reason why it can't be on your terms. It's a lot easier to defeat a clone currently dealing with a hangover or food poisoning. If you're patient enough to wait for the right moment (like after your clone goes on a hefty bender), your double will barely have the strength to get out of bed, let alone fight herself to the death.

- Make sure it's final. The most common mistake when doing battle with any arch-nemesis is to assume that you've won before the fight is really over. During a fight, you may injure your clone to such a degree that you think you have the upper hand. At this point, you may be tempted to flaunt your triumph with a rehearsed victory dance. Don't be foolish (also, don't be such an ass). You're merely giving your clone a chance to recover, and presenting an opening for a counterattack.

· Discretion is the better part of valor. If your clone is really working you over, the best strategy is to get out of there (fast) and give yourself a chance to regroup.

When planning a confrontation with your clone, physical combat should be your last resort. Tricking your clone into having dinner with your ex-girlfriend is funny. Punching your clone in the face is felony assault.

Covering your tracks

If the battle gets so out of control that your age-matched clone meets his demise at your hands, you're going to be in some serious trouble, and you're probably going to need to dispose of the evidence. If you're clever, you might be thinking, "I could make my clone's death look like my own suicide—then I wouldn't even need to hide the body!" But if you're *really* clever, you already know it's not that simple.

Staging a suicide typically requires a serious amount of planning and foresight. The "I killed myself" angle is a bit of a stretch if your clone has been pummeled to death with a Wiffle ball bat, no matter how convincing your fake farewell note. Even if you arrange the perfect "suicide" for your double, modern forensic science is still against you. Recall that differences in fingerprints, retinal patterns, or even dental history can easily distinguish your clone's body from your own. If the police or medical examiners have any reason to take a closer look, the jig will definitely be up.

While it's a bit morbid, the above strategy works best with a death that leaves little evidence and little to the imagination. For example, a "sleeping pill overdose" offers too much to be investigated and too many questions to be asked. Better to go with something more definitive—like jumping into an active volcano—that shuts the door on any postmortem examinations. Also, don't bequeath all of your clone's stuff to yourself in his or her fake will. You're smarter than that.

Honestly though, the fake suicide approach is probably more trouble than it's worth. If you simply need ideas for hiding a body, we recommend you watch *The Sopranos,* or any Scorsese gangster film of the last two decades.

Genetic preemptive strikes

If you're expecting conflict to arise between you and your clone, it might be a good idea to take out a little genetically engineered insurance. Though not technically clones, the bioengineered Replicants in the movie *Blade Runner* are limited by design to a four-year life span to prevent them from achieving enough personal growth to have any dangerous ideas about their status. Some of the Replicants are also implanted with false memories, which can be used, indirectly, to control them. A clone that remembers you attending all of the Little League games of her nonexistent childhood is likely to be fond of you . . . until the truth comes out. It doesn't matter how good you are at creating and implanting false memories if you can't live the lie. One too many contradictions between what you remember and what your clone remembers, and you're going to have a pissed-off emo clone to deal with.

While an emotionally dependent clone is a double-edged sword, a chemically dependent clone is equally vile yet much less effective. In *Jurassic Park,* the dinosaurs are genetically manipulated to be unable to synthesize the amino acid lysine. Without a steady source of lysine in their diet, they would quickly die, but the dinosaurs find edible plant life on the island with enough lysine to survive without the park's supplements. Their survival demonstrates a critical flaw in the control-by-chemical system: you're very rarely the only one with access to that chemical. A crafty clone will figure out how to feed the monkey you put on his back without you, and then come looking for revenge. In this case, we'll be rooting for the clone, because creating human beings that are artificially dependent on you is kind of evil (and ultimately pointless).

If you feel a little guilty about creating an alternate version of yourself with a bunch of genetic defects, there are moral alternatives for the canny (not to say paranoid) would-be cloner. Consider including a genetic tag to distinguish yourself (and your genetic material) from your clone if things go awry. An extra bit of noncoding DNA slipped into your clone's genome is all you need to prevent a lifetime of confusion; with genetic tagging, a simple blood test will be enough to prove your identity, and a properly chosen tag will be completely harmless to your clone. In the event that your carefully engineered progeny turns to a life of crime or attempts to displace you, it will be simple enough for the authorities to tell the two of you apart.

If genetic screening seems like too much trouble, introduce a genetic change that produces a visually distinguishable clone (we suggest eye color). Or, if an exact body double is actually desired, a small birthmark or tattoo in a private location could also provide a unique form of ID. In *The 6th Day*, each clone was marked with a series of small dots on the inner eyelid to denote that clone's unique generation number—a discreet but simple way to differentiate between genetic duplicates.

Clone versus clone versus clone versus . . .

If you're still struggling to defeat a single, age-matched clone by yourself, you may be tempted to try some unorthodox methods for ganging up on your genetic duplicate, like making another duplicate. Stop right there. This is a huge mistake. Even if you and your new clone can team up to beat your old clone, you'll be right back where you started, and the new clone will probably feel like you owe her. Also, what's to stop the new clone from joining forces with the old clone? They probably have more in common with each other anyway.

The only time we recommend the above strategy is if you use a little genetic foresight when making your newest double. A "good clone versus evil clone," or "brainwashed clone versus independent clone" scenario might actually work out in your favor (and could be *awesome* to watch). But "regular clone versus regular clone"? You're just asking for trouble. If you're looking to get trounced by multiple copies of yourself, why not skip ahead to "army of clones" right now and save yourself some time?

If you can't beat 'em, join 'em

When all else fails, try bargaining with your clone. Perhaps you can set up a visitation schedule for your friends and family, or divide your belongings equally and start the relationship anew. Besides, it really shouldn't be *that* hard to work out your differences, considering you're already genetically identical. In some cases, you may even find that your "evil" clone was actually created by sinister external forces. Teaming up could give you a chance to defeat the real villains.

Technically speaking, if you even have a clone, then you're a clone too—your clone's clone. The dude can't be that bad . . . right? If you and your clone(s) would just stop fighting each other for a few minutes, you might even have some fun together. Think of the possibilities!

- Cloning pranks. Exit a room through one door as your clone walks in through another. It's like *The Parent Trap*, only funny. Okay, it's still not that funny.
- Be your own wingman. This is great if you're a swinging bachelor, except that half the time you'll be stuck with the obnoxious best friend. If life has consistently taught us one thing, it's that twins are sexier than individual people, which, by deductive reasoning, suggests that clones will be the sexiest people alive.

- Instant substitutions in sports. This is a great way to conserve energy if you're an immoral athlete. Totally useless otherwise.
- Twenty-four hours of you. Rotate your sleeping schedules so that there's always someone "on the clock." This also makes it easier to share stuff with your clone if you don't have two of everything. Twelve hours of sleep ain't bad either.
- Olympic glory. Create the greatest synchronized swimming team ever assembled.

Of course, if your clone is insistent on being a douchebag, the above options don't apply. And if he's a *bioenhanced* douchebag, then you're really in hot water. Read on.

HOW TO DEFEAT A BIOENHANCED CLONE

An enhanced clone will retain the majority of the original's genetic code, but also possess genes that have been tweaked, added, or removed to make a clone "superior" to the old-and-busted version (i.e., you). Simple alterations in some muscle-related genes can give your clone an Olympian body without ever lifting a thumb. While an imperfect gene pool may have left you with poor eyesight, ample love handles, and an overall dearth of athletic ability, your fit and nimble clone is only a few nucleotide tweaks away. On the surface, your überdouble may look reasonably similar to you, but underneath is a bioenhanced body that provides advantages you won't ever match with diet and exercise alone.

The most obvious strategy when dealing with a superclone is to fight fire with fire—bioenhance yourself to one-up your clone's enhancements. There are two main ways to approach a "battle of the upgrades": (1) Over-engineer—if your clone has 20/20 vision, you should get telescopic eyeballs; if he can bench-press a small car, you should be lifting buses; (2) Design around—your clone may have the speed of a world-class sprinter, but she'll never be able to catch you if you can fly.

Bioenhancement one-upmanship can quickly turn into a genetic arms race between you and your clone—the biological equivalent of an absurd comic book duel where the two con- testants take turns trying to outgun each other, each revealing a larger and more ridiculous weapon than the last. Genetic engi- neering may eventually help you defeat your bioenhanced clone only to leave you looking like the Incredible Hulk on steroids.

Of course, without counterenhancements of your own, vic- tory is increasingly unlikely; but even if you lack the necessary resources to overhaul your existing frame, you may still have some natural advantages over your pumped-up alter ego. A lifetime of experiences may wear you down over the years, but it can also yield benefits that your upgraded clone hasn't had time to obtain. Immunity to certain diseases is acquired only through exposure to and recovery from various infectious agents. They say that chicken pox is harder on you as an adult, and your clone is one babysitting session away from the experi- ence, while your previous exposure has left you immune. Or, if your clone has an IQ of 200 but never went to college, he or she is probably unfamiliar with the old "beer before liquor, you've never been sicker" adage—even an Einsteinian intellect is no match for Bud Light and Jack Daniel's. Bioenhancement is great, but it won't necessarily replace years of preparation or decades of personal experience. The trick is to figure out what your clone lacks and then exploit those weaknesses.

If you're especially clever, a bioenhancement can be an exploitable weakness, particularly if that enhancement is a work in progress. For example, if your clone is a bioengineered supertaster—gifted with an unusually discriminating palate—he or she may be capable of identifying fine wine with a sip while you can only tell Pabst Blue Ribbon and Coors apart by the labels. On the other hand, the same clone may also be an absurdly picky eater and prone to gagging on food that's been adulterated by a pinch of garlic too faint for mere mortals to taste. Bioenhanced eardrums could have similar drawbacks. For some, an alarm clock is a mild nuisance; for a clone with unnaturally acute hearing, it's like a Megadeth concert cranked up to eleven. And while you might find your significant other's singing along to the car radio endearing, imperfect pitch is like fingernails on a chalkboard to your doppelgänger's tricked-out cochleas.

Not all fights are physical: a social victory can be equally vital in the battle against your next-gen double. In a bare-knuckle brawl, your puny birth body will be no match for your sinewy custom-engineered clone, but perhaps you're a lover, not a fighter. Fortunately for you, not everyone is attracted to the insanely muscle-bound. Your statuesque clone may seem like a shoo-in to steal your spouse and take over your life, but your significant other might actually admire your more vulnerable physique and find your duplicate's obvious musculature a little intimidating. In life and in love, flaws can define you, even if that definition is merely "not completely out of my league."

There's also a chance that a bioenhancement could yield unexpected side effects, especially if the genetic engineers aren't completely sure what they're messing with. A classic real-life example comes from sickle-cell anemia, a genetic blood disorder that requires two copies of a "faulty" hemoglobin gene to manifest itself. Repairing both copies would cure your clone's anemic condition, but would also leave the clone with lowered resistance to malaria—a strange and unexpected drawback. Much of our genome is still uncharted territory—we know what's out there, but we have no idea what it *does*. So who's to say what might happen when we start fiddling with genes that we don't fully understand? Your "enhanced" duplicate might also come with a fair share of glitches if the biological designers aren't careful. Be on the lookout for anything that could be used against your upgraded double.

An ultraenhanced clone with no biological weaknesses at all will probably be pretty cocky, and that alone can work to your advantage if you're clever—even Muhammad Ali got knocked down from time to time. Note, however, that Ali never got knocked *out*, so if you do stun your bioenhanced clone, be sure to get in a few good kicks. Right in those genetically chiseled abs.

HOW TO DEFEAT AN ARMY OF CLONES

Dealing with a standard clone is tricky. Dealing with an enhanced clone is even trickier. But dealing with a hundred clones? Or a thousand? Unless your name is Neo and you're fighting a playground full of Agent Smiths, you're probably in trouble. The previous clone-defeating techniques in this chapter still apply, but now on a much larger scale.

The best way to defeat an army of clones is to make sure you're in control from the get-go, because really, the only reason you should ever have an army of clones is to rule over them. When designing a clone army, be sure to include some genetic preemptive strikes (the more the better), like programmed obedience or a carefully designed biological fail-safe. We certainly don't recommend you start making multiple versions of yourself, but if we can't stop you, at least make sure you're in the driver's seat.

Remember, an engineered clone army can be a mixed bag. In *Star Wars: Attack of the Clones,* the troopers led by Jedi Knights are genetically programmed to be highly disciplined and, more important, obedient. The end result? A big pile of dead, betrayed Jedi.

With or without genetic indoctrination, few people covet an army where one soldier can't tell the other soldiers apart, and where rank is challenged by the knowledge that a superior officer has nothing that you don't have. Working closely with several copies of yourself would be trying for even the most patient of us—no one can push your buttons like family, particularly when that "family" consists of ten thousand identical twins. If faced with this unlikely scenario, be prepared to take advantage of your inability to get along with yourself.

With each clone, from the first to the fiftieth, you forfeit increasing degrees of control. If your clone decides to start making bootleg copies of himself, there's very little you can do to prevent this. Luckily, an unfriendly army that looks exactly like you presents an unparalleled opportunity for infiltration, especially since you will also have a unique glimpse into their psychology as you and they will share many of the same characteristics—both strengths and weaknesses—even if you're only *genetically* identical. Direct confrontation with an army of clones is usually a bad idea, if for no other reason than the fact that you're outnumbered. But by masquerading as one of the masses, you may be able to sabotage them from the inside, or at least keep yourself alive until you figure out a better strategy.

Your last option is to beat the clones at their own game: fight the army with another army. But depending on the size of the clone battalion, you may not have enough Facebook friends to make it a fair fight. If your clone army is really getting out of hand, you'll probably need to employ some government or law-enforcement aid as a last resort. You're going to have a lot of questions to answer once the police round up the last of your rogue duplicates, but those questions will be less grueling than trying to outmuscle yourself a hundred times over.

We hope you'll never be faced with an unrelenting horde of yourself, but if you are, we suggest ingenuity and resourcefulness over brute force.

HOW TO DEFEAT SOMEONE ELSE'S CLONE

As the biotech revolution unfolds, it's highly unlikely that you'll be the only one with a clone. In fact, as the technology progresses, cloning could easily become available to any Tom,

Dick, or Mary with a double helix and a credit card. Without a little discretion, we could be in for some bumpy rides. We all have that friend from high school who could use a little help in the common-sense department, and he'll probably be the first one to get himself cloned on a dare or a whim. Sure, it's not exactly your problem, but what happens when your careless chum is replaced by a group of rowdy young doubles who aren't willing to return the circular saw their original borrowed from you last June?

When it comes to defeating someone else's clone, you may find yourself in a unique position to help identify the original or spot the impostor. Remember the lessons from Chapter 2 and be cautious when dealing with a potential faker. The most important thing is, use your best judgment and avoid any rash decisions. Mistaken identity works both ways when dealing with a clone and his original, and the last thing you want is to be the guy who accidentally hands poor Jimmy over to the police while his double has brunch with Jimmy's wife and kids. You may be the only person left who can positively identify the impostor, so your choices will make you either a hero or a bonehead.

It's likely that the new clone will try to get you on her side—either by impersonating the original or by outclassing her—but there's also a possibility that she'll perceive you as a threat instead. Give the new clone any reason to think you're working against her and you may suddenly find yourself in the line of fire. This can be particularly confusing if you're still unsure who the real clone is. Be sure to feign ignorance until you get the facts straight.

Once you're certain that your friend's clone isn't trying to eliminate you, too, consider using yourself as a decoy to lure the impostor. The clone may think you think she's the original, but you know that she doesn't know you know she's a faker. While she thinks she knows what you're thinking, you actually know she has no idea, you know? It's the old "double-cross" strategy, and you're right in the middle. Just be sure you know which side you're playing.

Even if you can't fool the clone with your clever mind games, you can still help beat her at the numbers game. The more allies your friend has, the better, and she'll be counting on you for support. A single clone stands little chance against a barrage of cousins or co-workers.

In addition, try not to be swayed by an "improved" clone. This new version may seem superior at first glance, but he is not the same person you used to play softball with. Even if he "remembers" all those fun times you had together, he was never actually there for any of them. Doesn't the original deserve a little credit? Besides, even if you never really liked the old Jimmy, he still has a right to keep his identity and his life. Or, if Jimmy's just being a prick, perhaps a little "clone intervention" is in order. Either way, you'd expect the same help in return from your friends and family, wouldn't you?

When the clones arrive, be sure to stay sharp, particularly as the technology advances. Ask yourself: Has your lifelong friend suddenly lost a few pounds? Are her teeth mysteriously whiter than usual? Where did she get that new blouse? Maybe you're just being paranoid. But then again . . .

HOW TO DETERMINE IF YOU'RE A CLONE

One last word of caution before challenging your clone to a showdown: Have you considered the possibility that *you're* the clone? In many cases, it will be pretty obvious that you've been manufactured as someone else's duplicate (say, if you live in a research facility and all your "friends" wear glasses and white lab coats); other times your status might not be so clear (if you've been programmed to think you're the original). Even if you can remember more than the last three weeks of your life, be wary of falsely implanted memories; possessing a brain full of childhood flashbacks is a little suspect if you can't remember your own address.

A true "mind-clone," however, would share all of the memories of the original but could still distinguish himself via physical differences. Do you have any distinguishing marks (tattoos, scars, wrinkles, etc.)? If not, why not? Even supermodels need extensive airbrushing to achieve those magazine-ready looks, because no one has a perfectly flawless complexion. If you're a forty-year-old man with the silky-smooth skin of an infant, this could be the result of having spent most of your life in an artificial uterus (deal with it). Without arduous biological fine-tuning, a clone and her counterpart will likely have some small but fundamental genetic differences as well—in the mitochondrial DNA, for example. With some very clever (and very specific) genetic testing, a clone could potentially be distinguished from his or her genetic forebear.

If you eventually discover that you *are* the clone, you may still want to "defeat your own original" anyway. Nevertheless, you should at least get the semantics straight before you start.

CONCLUSION: YOU ALWAYS KNEW YOU SHOULDN'T HAVE TAKEN YOURSELF SO LIGHTLY

Make no mistake: A clone will be an adversary to be reckoned with, *if* that clone is your adversary. Act honorably and chances are you two will never come to blows. Create another human being solely for personal gain and you're breeding trouble. If you choose to dabble in cloning, know exactly what you're getting into, and be prepared to live with (and get over) yourself.

EPILOGUE

Cloning may be good and it may be bad. Probably, it's a bit of both. The question must not be greeted with reflex hysteria but decided quietly, soberly, and on its own merits. We need less emotion and more thought. —RICHARD DAWKINS

When it comes down to it, "defeating" your clone might not actually be the best solution, though it does make for a sexy title. If this book had been called *How to Befriend Your Own Clone: And Other Tips for Enjoying the Biotech Revolution*, would you still have picked it up? Impending doom tends to generate interest—the problem is that societies tend to get carried away by fictions and opinions that have tenuous connections to reality; the answer is to approach such propaganda (unwitting or otherwise) with a healthy dose of skepticism. A culture fed a steady diet of alarmism risks breaking out the torches and the pitchforks without sober consideration.

Reproductive cloning, human bioenhancement, and artificial organisms may sound today like science fiction, but what is thought of as impossible by one generation is often taken for granted by the next. Despite the kerfuffle that in vitro fertilization caused when it was an emerging technology, now the phrase "test-tube baby" seems rather quaint. Yesterday's outrage is no indication of what tomorrow's will look like.

While biotechnological progress is unlikely to halt or end the world, new solutions to old problems often lead to new problems. The trick is to separate the likely problems from the impossible or the ridiculous. We have a limited amount of attention to pay to the future; no worry should be wasted that could be better utilized. With a little forethought, defeating your own clone should be totally unnecessary, but we've tried to prepare you for every contingency, just in case.

ACKNOWLEDGMENTS

We would like to thank Mairi Beacon, Kevin Cheng, Wendy Doyle, Steve Frishcozy, Karen Healey, Theresa Isidro, Randall Janairo, Wesley Keppel-Henry, Anne Kim, Natalie Mitchell, Pam Reynolds, Ray Schmidt, and An-Chi Tsou for their generous and helpful comments.

Thanks also to everyone at Bantam Dell, especially our editor, Philip Rappaport. His insight and hard work improved this book immeasurably. (Seriously, we've tried to measure it.) Ming Doyle's illustrations obviously didn't hurt, either.

Last, we would like to thank our agent, Laurie Fox, without whom this book would have died not far past the "drunken idea" stage. We are a very lucky pair of first-time authors.

KYLE KURPINSKI holds a Ph.D. in bioengineering from the joint graduate group between the University of California, Berkeley and the University of California, San Francisco. He has worked in various labs across the country including the National Human Genome Research Institute in Maryland, the Brookhaven National Laboratory in New York, and the Lawrence Berkeley National Laboratory in California. Kyle grew up in the suburbs of Detroit and attended the University of Michigan, Ann Arbor, where he received his bachelor and master's degrees in engineering. His original goal in life was to become either a neurosurgeon or a stuntman, but he put these dreams on hold to pursue the jet-setting lifestyle of a research scientist. After college, Kyle fled to the West Coast in search of warmer climates and better drivers. His quest remains only half successful.

Kyle currently works for a biotech company in the San Francisco Bay Area where he develops new technologies for tissue regeneration. This may sound impressive on paper, but Kyle is the first to admit that the majority of his time is spent transferring small amounts of liquid between various containers.

TERRY D. JOHNSON is a lecturer in the bioengineering department at the University of California, Berkeley. He began his lecturing career in the chemical engineering department at MIT. The subjects of his classes range from tissue engineering to mathematical biology, displaying a versatility that has prevented him from achieving any actual expertise in a single subject.

Terry is a machine that takes in caffeine and alcohol and outputs hair, paralyzing self-reproach, and the occasional PowerPoint slide.